A BEAUTIFUL OBSESSION

A BEAUTIFUL OBSESSION

JIMI BLAKE'S WORLD OF PLANTS AT HUNTING BROOK GARDENS

JIMI BLAKE AND NOEL KINGSBURY

filbert *press*

First published in 2019 by Filbert Press
filbertpress.com

Photography credits appear on page 221

A catalogue record for this book is available from the British Library.

ISBN: 978-1-7399039-6-1

10 9 8 7 6 5 4 3 2 1

Designed by Studio Noel
Illustrations by Hello Marine/meiklejohn.co.uk
Printed in China

TO MY LOVELY MOTHER KATHLEEN –
MY FIRST INSPIRATION. THANK YOU
FOR ALL YOUR LOVE, SUPPORT
AND ENCOURAGEMENT

JB

CONTENTS

PART 1

EARLY BEGINNINGS

PART 2

A WALK AROUND THE GARDEN

PART 3

PLANTS AND PLANTING

FOREWORD

Jimi Blake is an exciting young man. He combines a deep love and inquisitiveness for plants with an artistry and adventurousness that is a joy. His garden at Hunting Brook is endlessly full of surprises, a dynamic canvas on which Jimi creates. New plants come and go, and the garden never stands still, and each visit leaves you inspired for more. Jimi has had various influences in his gardening life – but today he clearly stands very much as his own man. His appetite for all plants is consummate – vacillating from the delicate and curious annuals suitable for weaving, to the bold and muscular sub-tropicals that punctuate Hunting Brook. Jimi has a sensitive hand but also one that delivers bold strokes, his style is rich like an opulent jewel box. And with Noel Kingsbury's astute insight into a gardening life, this book is a fascinating read to all of us who tread the same path.

—Fergus Garrett
Great Dixter, April 2019

PART 1

EARLY BEGINNINGS

WELCOME TO HUNTING BROOK

Gardening, for me, truly is a beautiful obsession. I feel blessed to have had it from a very young age and it becomes more profound as each year passes. It is a privilege to be the guardian and curator of this sacred land at Hunting Brook, which for me has become a plant canvas where I can experiment with, grow and collect plants from all over the world. Hunting Brook Gardens is a place of creativity – it challenges me to make new areas within the garden, try new combinations and push my own concepts and knowledge of plants. Visitors talk of experiencing a special feeling when they walk through our gates, as if being transported into another world, and I really hope this beautiful book captures something of that and inspires you to make the journey to Hunting Brook someday.

—Jimi Blake

It is sometimes said that walking around a garden should take you on a journey. In the case of Hunting Brook, the garden which Jimi Blake has been making in Co. Wicklow in Ireland since 2002, this is very much the case. Open to the public as a business, Jimi's garden has been designed to take its visitors through a variety of terrains. For those expecting just a garden it will be a surprise, for the journey will take them into places which are well beyond what we normally think of as garden spaces. The garden and landscape here offer a broader prospect, through an extraordinarily wide range of plant life and ideas about how plants can be put together. In Part 1 and Part 2 of this book I'm going to take you on a walk round the garden with Jimi at my side and tell you something of Jimi's personal journey. You'll hear from Jimi on the feature spreads and in Part 3 where Jimi describes the practicalities involved in maintaining the garden and shares his current favourite plants.

—Noel Kingsbury

CO-AUTHORS JIMI AND NOEL UNITED IN THEIR LOVE OF FLOWERY SHIRTS

PREVIOUS PAGE
The sparse foliage of *Pseudopanax lessonii* 'Tuatara' (left) and *Olearia lacunosa* (right) frame silver *Parolinia ornata* and dark red *Salvia* 'Love and Wishes'.

LEFT
Ashley's Garden was designed and planted in 2012. It was originally the old car park and is named after Jimi's friend Ashley who loved the garden.

ABOVE
Hunting Brook Gardens is in the bottom left corner of the picture and in the distance are the Wicklow Mountains.

Jimi Blake has been a lifelong gardener, starting out indeed with his fingers in the soil of this very land, for the 8 hectares (20 acres) of Hunting Brook are the result of a share-out of the former family farm, Tinode. Born in 1971, he gardened here as a youngster, moved to Dublin, and then in 2001 returned to build a career as a transformer of this up-and-down patch of land into an experimental and educational garden, a place to hold courses in gardening, yoga, meditation and healthy living, and as a base for a surprising amount of travelling, for Jimi now runs garden tours and has an increasingly busy overseas lecturing schedule. Central to his educational mission is the Plantsperson's Course, which takes place one day a month for ten months of the year. At the time of writing, this course has been running for 12 years and has had around 300 students; it has undoubtedly helped to educate a whole new generation of garden enthusiasts in Ireland.

Jimi's mother is a keen gardener, and it is pretty clear that's where the passion comes from. His sister June is also a notable gardener, having run a nursery and now herself being in possession of a fine garden that's open to the public, just on the other side of the hill to Jimi. Her name is known far and wide through a particularly fine yellow polyanthus primula she bred, 'June Blake'.

Once the traveller is free of the outer suburbs of Dublin, 'real' countryside is reached quite suddenly, with hilly roads winding through lush green fields and trees. The sign for Hunting Brook takes you up a fairly steep lane with woods on one side and pastureland on the other, very typical of Ireland or indeed of anywhere in the western part of the British Isles. That you have arrived is indicated by the kind of funky galvanized gatepost and sign that stands out in this largely conservative countryside, a bit like a Californian relative at a country wedding. If you visit in August, a froth of *Persicaria campanulata*, running riot on the bank opposite the entrance, hints at a plantsman's lair.

After passing through the entrance you are directed down to a car park, from where there is a walk upwards. Almost the first thing you notice are some rather unusual trees on the bank to the left, behind which is a wooden house. The banks – there are actually two, one on each side of the walk up into the garden – are exuberantly planted, but it is the trees which really stand out. They are medium-sized, 8–10m (26–33ft) or so tall, with upright-swept branches, and really do make you stop and look, for two reasons: one is that they have very few branches and the other is that their leaves are twice pinnate, so they divide once and then again. This is a combination of characteristics that makes them very distinct from the trees we are familiar with in the northern temperate part of the world. They cast very little shade, so not surprisingly there is a great deal growing beneath, and a lot of that looks unfamiliar. Clearly you are somewhere special.

I have been visiting Jimi for some six years now, and I have seen plenty of changes. In particular, this first part of the garden, that to the east and north of the house, was once dominated by perennials and grasses. It is now more exuberant, with more bright late summer flowers, particularly dahlias and salvias, and a good many exotic-looking foliage plants that at first sight are impossible to identify. Radical changes are a major part of garden life here, for Jimi has a restless love of new plants and new ways of putting them together. Many gardeners replant, replan and rejig until they reach some sort of satisfaction, then leave be. Jimi does not do this. I feel that for him satisfaction is somehow boring; that a successful planting should not be allowed to rest on its laurels for too long before being dug up and replaced with something that is testing new waters.

Before we go any further, we need to take stock of what the climate is like. Seeing all the colour, barely hardy plants including Mexican salvias and exotic foliage such as bananas, one might think, 'Oh, this is Ireland, they can do that here.' In fact, Hunting Brook is, Jimi reckons, 'One of the coldest gardens in Ireland, we are on the edge of the Wicklow Mountains.' The garden is at an elevation of nearly 305m (1000ft) and the most tender plants do have to be taken inside in the winter. This makes Jimi's garden a very useful and realistic model, showing what can be done to create a 'hot' look in conditions not dissimilar to that of many gardens broadly at this latitude. A balmy Gulf Stream garden this is most definitely not! In fact the cool climate is probably exacerbated by the soil, an acid clay which, like all heavy soils, is slow to warm in the spring.

This exuberant 'front of house' area is only a small part of Hunting Brook. The rest moves at a slower pace, although much is still very experimental. There is a woodland area with a wide variety of mostly spring-flowering plants, a deep valley in the shade of an old plantation of mixed coniferous and deciduous trees, and then an area of grassland. This whole area, less obviously gardened, is slowly being planted up and will one day almost certainly be Jimi's greatest legacy, for this is where he is doing some important long-term planting, in many cases with species we have very little experience of in cultivation.

So, in this book we are about to start on a journey through a visionary and experimental garden and landscape. However, there is more to it than the horticulture. As Jimi describes it, 'Whatever piece of land you have can help you stay grounded, keep you connected to mother earth and bring about deep healing. . .this place is not just about providing me with an income, about doing more courses or getting more people through the gate – I want Hunting Brook to make a difference to people as well. The healing side has always been a big part of what this garden is about, because it's been a big part of me.'

Jimi has always run retreat days, catering with his own vegetarian food. Occasionally he hosts groups with very specific needs. He says, 'I had an amazing group of young people in recovery from drug and alcohol problems here – there's an organization that brings them out from Dublin to do wild camping. I went to see them in the morning and it had been an amazing experience for many of them, who had probably not been out of the city since they were kids. Their faces were lit up from being here, and they were looking around at everything.'

A horticultural extravaganza but also a spiritual and healing place; welcome to Hunting Brook Gardens.

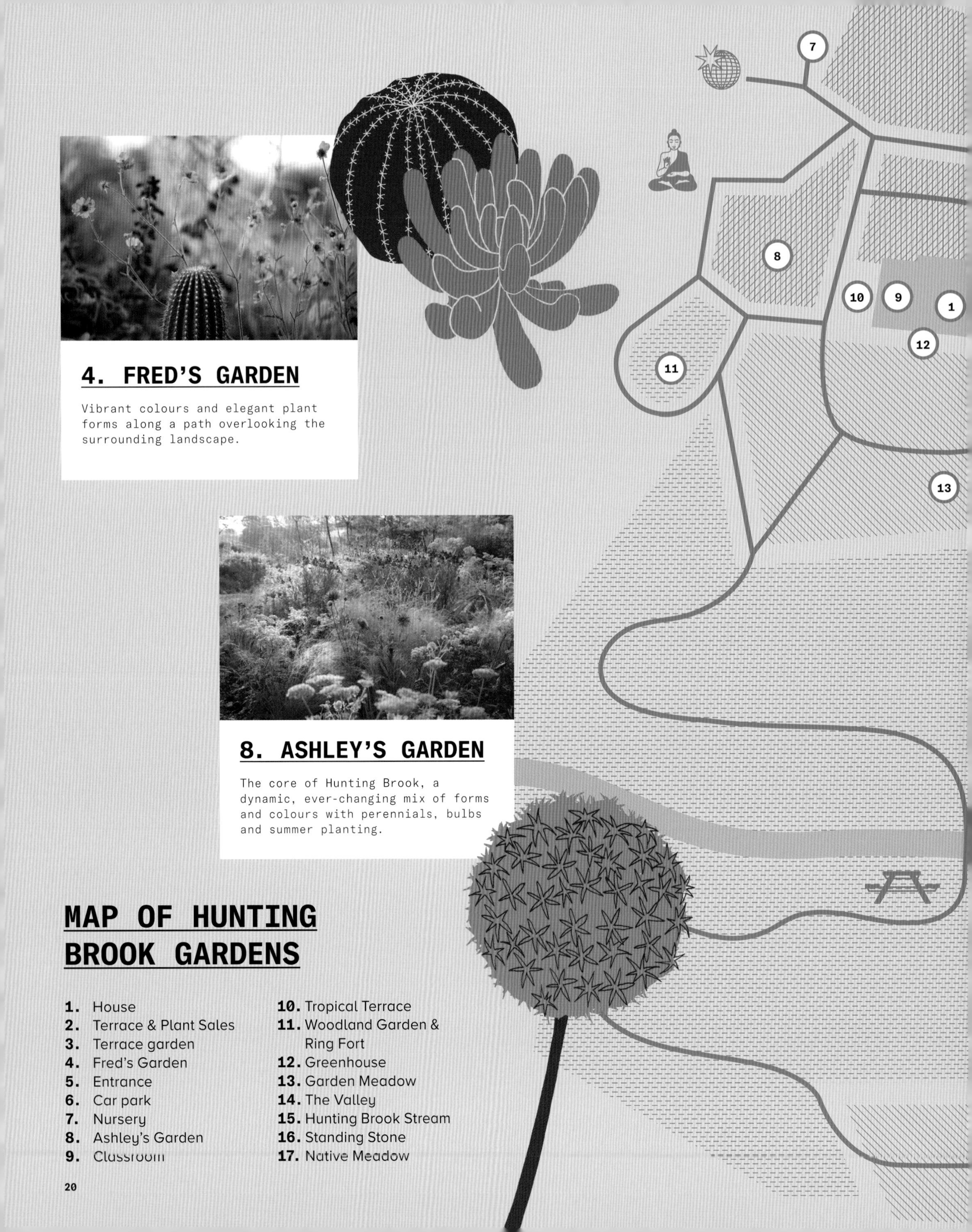

4. FRED'S GARDEN

Vibrant colours and elegant plant forms along a path overlooking the surrounding landscape.

8. ASHLEY'S GARDEN

The core of Hunting Brook, a dynamic, ever-changing mix of forms and colours with perennials, bulbs and summer planting.

MAP OF HUNTING BROOK GARDENS

1. House
2. Terrace & Plant Sales
3. Terrace garden
4. Fred's Garden
5. Entrance
6. Car park
7. Nursery
8. Ashley's Garden
9. Classroom
10. Tropical Terrace
11. Woodland Garden & Ring Fort
12. Greenhouse
13. Garden Meadow
14. The Valley
15. Hunting Brook Stream
16. Standing Stone
17. Native Meadow

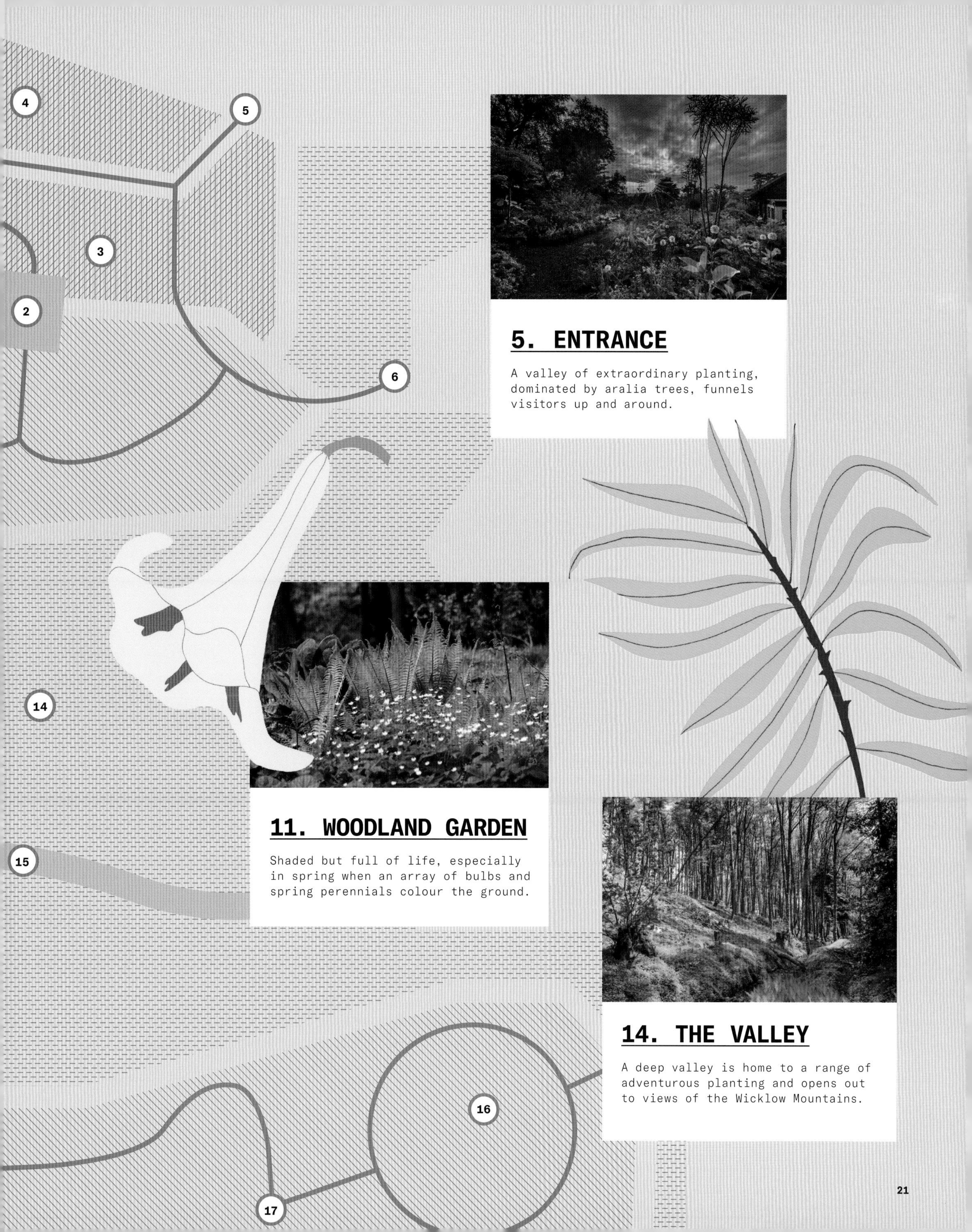

5. ENTRANCE

A valley of extraordinary planting, dominated by aralia trees, funnels visitors up and around.

11. WOODLAND GARDEN

Shaded but full of life, especially in spring when an array of bulbs and spring perennials colour the ground.

14. THE VALLEY

A deep valley is home to a range of adventurous planting and opens out to views of the Wicklow Mountains.

G-STAR

A BEAUTIFUL OBSESSION

Nearly all the gardeners I interview took to that way of life because someone in their family encouraged them, or maybe just set an example. Jimi is no exception. In his case it was his mother, Kathleen Blake. 'It was a family farm – my dad, Jim Blake, was a farmer and butcher with sheep and cattle,' Jimi recalls. 'There was a big garden, mostly rhododendrons. My family bought the place 60 years ago, when it was completely overgrown and the house was a shell. My mother moved into the middle of nowhere when she was only 23, with three kids.'

Tinode House had once been a grand house, built by a wealthy landlord who was, unusually, from a Catholic family. June says, 'It passed to an English colonel, and then in 1923 it was torched by the Anti-Treaty forces during the dying days of the Irish Civil War. The stables, servant's quarters, a little oratory, a chapel, and a billiard room were all that were left, and it was in them that we lived.'

Like many of the time, the Blakes were a big family; they had five children and after a gap of 12 years, when Kathleeen was 42, Jimi was born. As a result, he must have had more of his mother's time than any of his siblings; crucially, he notes also that, 'She had a bit more time to garden with only me around and so I was always with her out in the garden.' Of his siblings, June is a distinguished garden designer in her own right and has a highly renowned garden, very different from her brother's a little way up the road. Mary, an archaeologist, lives in the Burren in Co. Clare, one of the few areas of exposed limestone in Ireland and famous for its diverse and spectacular vegetation; part of her work is taking groups around the archaeological sites and wildflower locations of this most special of landscapes. Gerard is passionate about trees, while Phyll, who lives behind Hunting Brook, is a wonderful vegetable gardener.

Kathleen Blake gardened from very early days at Tinode, mostly vegetables to feed a growing family, but as time went on she added more ornamental plants. 'My mother used to grow veg; I'd be left in the pram at the end of the row she'd be working on,' Jimi says. 'I suppose I started at age five or six.' A 'room full of cactus' came later on, and he recalls getting plenty of support from an early age: 'My mother bought me a polytunnel when I was very young, and a glasshouse.' Kathleen remembers him getting the polytunnel. 'He looked like he'd won the sweepstakes,' she says.

June, 16 years older than Jimi, was not around for much of Jimi's childhood, but recalls him 'Always being with our mother, who was always in the garden, and he was always gardening. Some of his first words were to do with the garden. We used to tease him, because "polyanthus" was one of the first words he said, but of course there were a lot of names he couldn't say properly as a child, "nasturtium" being one of them, and in fact he still can't say it properly!' As with many of the gardens around the big houses of Ireland, there were masses of rhododendrons, grown large and luxuriant with the decades. 'We all played in them,' says Jimi, and indeed, rhododendrons are a wonderful place for children; their dark enclosing foliage shuts out the adult world and the twisting branches offer endless possibilities for climbing and all sorts of imaginative play.

Now sold, the family home can just be seen above June's garden, its façade almost hidden by trees, which include some enormous specimens of *Rhododendron arboretum,* the red-flowered species which can grow to 20m (66ft), making it the tallest-growing of the genus. As if they were not enough, Jimi recalls, 'We propagated them too, I was very young when we were doing that, both air-layering them and layering them in the ground.' Air-layering is a fairly sophisticated technique for propagating plants, involving wrapping moss enclosed in plastic around stems in order to get them to root; that Kathleen used this technique says a lot for her gardening knowledge and confidence.

The family was clearly an entrepreneurial one, up for tackling various schemes. 'We did everything then,' Jimi says. 'We grew flowers in a polytunnel then dried them by hanging them from the rafters in the loft, which was huge. Then we made them into arrangements that were used in my sister Mary's hotel.'

As well as rhododendrons, there were perennial beds and a big rockery which would have been part of the original garden. Propagating plants was clearly something of a family obsession; Jimi says, 'My father liked propagating too and he would plant things out all the time, escallonias and other plants that would drive my mother mad.' Jimi vividly remembers as a child the feeling of excitement taking cuttings and potting plants up: '*Lonicera pileata* may be a common plant, but for me then it was a thrill as it was so easy to propagate,' he says.

When Jimi went to secondary school at the age of 12, however, his life changed. Secondary school can be a very difficult time for children of a sensitive disposition; in Jimi's case this was made worse by the school being run by the Christian Brothers, a monastic order. His problems at school were exacerbated by his protruding lower jaw, which led to many incidents of bullying. 'This needed to be corrected by surgery in the summer of my third year in school and I had to return the following September with my mouth braced . . . this caused me great distress at the time.' Another issue was 'trying to come to terms with the fact that I realized I was gay in the middle of secondary school, not easy anyway and the Christian Brothers and the fact that it was illegal in Ireland at the time only added to the anxiety.'

OPPOSITE
The main house at Tinode was a shell until the early 1980s when it was restored by Jimi's father. Jimi and his siblings spent their childhood playing in the ruins.

BELOW
An ariel view of Tinode after it was restored surrounded by huge rhododendrons. These were the gardens where Jimi started gardening.

One escape for Jimi, and perhaps the most important part of his education, was spending time with plants. 'I started working in a local garden centre, Kinsella's, at the age of 12,' he says, 'and I worked in a herb farm for a summer.' He grew and propagated many plants and loved keeping records of what grew and what didn't; in the polytunnel he used to plant all the cuttings in circles and then cover them in plastic, making mini-tunnels. He says, 'Working in the garden centre, I would take cuttings of whatever I could lay my hands on, mostly shrubs and trees, just ordinary shrubs, but they were new to me so they weren't ordinary.' The experience of working led Jimi to his first garden entrepreneurship, setting up a little shop down on the main road in the days before regulations on planning and health and safety came into being.

A true farmer's son, Jimi also kept sheep as a teenager; he owned 30–40 ewes and was adamant he was paid for them. Eventually, when he was 18, the earnings from the sheep and garden work gave him independence and he could move out of the family house. A home of his own was no doubt part of working out how he could live his own independent life.

Leaving home and school was his big break. 'I knew for years that I wanted to train in the Dublin Botanic Gardens,' Jimi says, 'so I went there straight after school. I loved it, absolutely loved it.' This institution is the National Botanic Gardens of Ireland, based in the suburb of Glasnevin on the northern side of the city. 'I did three years at Glasnevin,' Jimi says, 'although the course was run by Teagasc College, (pronounced 'chagas') an institution separate from the botanic garden. I was doing a Diploma in Amenity Horticulture. We worked in the different areas of the garden with so many things I had never seen before . . . the woodland plants, that was a really big thing for me, and I loved working in the glasshouses, behind the scenes, all the exotic stuff – at that stage everything was new to me so it was all really exciting. I was renting a flat at the time, which was crammed with things grown from cuttings I had swiped from the garden, and we'd get plants out of the dump at the Botanics.'

Jimi thinks he was a 'complete plant nerd at college.' 'There were nearly 50 of us, and a lot who really shared a particularly strong passion,' he says. One of them, another well-known name in Irish gardening, was Seamus O'Brien, now Head Gardener at the National Botanic Gardens at Kilmacurragh, Co. Wicklow. Jimi says they were definitely kindred spirits, while Seamus remembers one occasion 'when we bumped into each other outside the General Post Office in Dublin, we both had rucksacks full of perennial divisions . . . we were both fanatical about plants'.

Practical horticulture courses at the National Botanic Gardens always involve work placements, which in Jimi's case 'were often menial, like we had to go and work in the parks. My first placement was in Phoenix Park, raking leaves; the next one there was a choice of picking stones, picking litter or planting the hedges in the middle of the motorway around Dublin. I had another one that was so bad I requested a transfer, and I ended up at a nursery, St Anne's Park in Raheny, run by Dublin City Council. That was when I really got into propagating. There was a really nice guy called Ed Bowden running it, and he was huge influence on me at that stage.'

A JOB AT AIRFIELD

After college, Jimi had an extraordinary break, and like many such breaks it involved being the right person at the right time in the right place. In 1991 a position came up at Airfield, an estate farm on the southern side of Dublin. This was a project run by a charitable trust set up in 1974 by the two sisters who had inherited the property, Letitia and Naomi Overend. Jimi describes it as a 'farm and garden, quite out of control – it hadn't been properly cared for in years. It had been owned by the two elderly sisters. I think the gardener was in his eighties – maybe he was in his seventies, but to me he seemed really old.' The trust's aim was to turn the 15 hectare (38 acre) farm into a visitor destination centred on the production and promotion of local food, along with having some areas of ornamental planting. It needed someone with a combination of practical skills and vision, and at the age of 21, Jimi managed to land the job. He says, 'I would have had good references. The man running the trust was Edmond Kenney – he was a great man. I can still remember a phone call to him – I was standing, shaking, in a phone box in Dublin, an old-fashioned phone box of course. He told me to come up for an interview. It was just perfect. I had started working on private gardens for people after I left college and I really didn't like it.'

The opportunity offered to Jimi was quite exceptional. 'In general I had a free rein – I was head gardener,' he says. 'It was a big project and there was very little restriction on me. I was able to go mad propagating and buying plants, and I was able to start travelling. I made that part of my budget, that I needed to travel to educate myself so I could build the place up.' He recalls there were 'perennial beds covered in couch grass, some remains of good plants, some great old glasshouses, but it was a slow process as these renovation projects often are. Slowly but surely we progressed – we relied on volunteers and I also brought in students to help in the early years until I had more staff.' Key to Jimi's experience at Airfield was the opportunity to collect a wide range of perennials and grasses and all the other cutting-edge plants of the time. Jimi says, 'I was remarkably lucky in being able to experiment, just try things out. I had an amazing boss at Airfield – Maggie Giro. She advised me to go and visit Great Dixter, which we did together.'

OPPOSITE
Jimi working in the garden at Tinode.

ABOVE LEFT
Jimi and his mum at Airfield when he was head gardener. This is a raised border planted with a hedge of dierama.

ABOVE RIGHT
Jimi with Fergus Garrett when he visited Hunting Brook.

ABOVE
Jimi working in Airfield's walled garden where he completely redesigned the borders.

Great Dixter is the Sussex garden established by Christopher Lloyd (1921–2006), who was one of Britain's foremost writers on gardening for many decades. His legacy has been managed and built on by Head Gardener Fergus Garrett; Great Dixter now has one of the highest reputations of all English gardens, largely for an idiosyncratic planting style which relies for its impact on visual richness, high contrast and maximizing interest in any given area for as long as possible. Jimi remembers, 'I spent a day with Fergus, sitting on the grass with him, chatting about how to open a garden to the public. He told me that I should be the gardener and not do the tours taking people round because it takes up all your time; gardening is more important. Maggie spent the day with the people in the office to see how they ran it as a business. It was such a useful exercise to see how Fergus ran the volunteers and the students – that was brilliant.'

Jimi stayed at Airfield for 11 years and the whole experience was clearly quite a formative one. He says, 'That initial job you get when you leave college can really change your life – it certainly paved the way for the rest of my life, I think.'

Airfield's location in the city of Dublin meant that Jimi had the ultimate experience of combining the green and the urban. He started to go to every lecture he could at all the different horticultural societies and conferences. His main concern at Airfield was, he says, 'Plant collecting – the obsession with building up a collection. The trust didn't really know what I was doing, so it was like having my own garden in the city to play with.' The fact-finding mission to Great Dixter with his boss started a trend, he says: 'I was making a few trips a year to England then. I brought in lots of plants and got to know people in the UK nurseries.' These links have been crucial to Jimi's progress and success as a plantsman and garden-maker ever since.

Another key aspect of Jimi's work at Airfield was his first dipping a toe into educating other gardeners. 'I ran my own garden courses while I was there. I was lucky to be able to do that – it did not come naturally to me, standing in front of people and talking, but I needed to make a living and get my name out there and now I love it.' He also started teaching beginners' gardening courses on Saturday mornings.

In Ireland, the 1990s to 2000s was an era in which there was a distinct upsurge of interest in gardening. This was the time when Jim Reynolds' Butterstream was one of the most talked-about gardens and also when the inimitable Helen Dillon started her career. Helen first came to fame with a book co-authored with Sybil Connolly, *In an Irish Garden* (Weidenfeld & Nicolson, 1986). Connolly (1921–98) was a fashion designer who made a career out of working with Irish-produced textiles. The Rockefellers and Jackie Kennedy were clients, and it was Sybil's contacts that helped to get Helen's career as a garden lecturer off to a good start across the Atlantic. Helen went on to write, lecture and present the country's main gardening programme 'Garden Heaven' on RTE (the national television station) which was filmed at Airfield and this was where she first met Jimi.

'Jimi's his own man,' says Helen Dillon, emphatically. 'He is a true original – I've never known him copy anybody. But his focus is the plants; he's not interested in structure.' Jimi's site, being all woodland and slope, is one which is singularly not amenable to a strong design; sister June, with her much stronger design focus, has ended up with the more level, open site where vistas and axes and perspective are simply more relevant. 'He manages to do the magic, entirely without structure,' Helen says, 'quite extraordinarily so.'

'Helen's been one of the biggest gardening influences on me,' says Jimi, 'I have learnt so much from her, and she's given me a lot of plants. Years ago she did a night class in Dublin which I went to. I still look at those notes – she's very generous with information. She has an incredible appetite for new plants, and her eye for plants is amazing. I like her bravery in just getting rid of plants that aren't pulling their weight.' Indeed, Helen is famous for having got rid of what was regarded as one of the best lawns in Ireland, the pride and joy of her husband, Val, choosing to replace it with limestone paving and a canal instead.

'The first lecture I went to on modern planting was by James van Sweden, in February 1997,' Jimi says, 'and I remember I nearly blew up with excitement. It was the Irish Garden and Landscape Designers Association seminar. I remember coming back to Airfield and doing a new planting in that block planting style, which I had never seen before.' The partnership of James van Sweden, a Dutch-heritage architect turned landscape architect, and Wolfgang Oehme, a German garden designer and plantsman, had been extremely influential in the USA, effectively introducing perennials and grasses to public and other large landscapes. Their style was noted for its extensive use of monocultural block plantings, a practice now largely discontinued in favour of more complex perennial groupings, but it was probably their confident promotion of grasses that brought them the attention they received in Europe.

Around the time of the Millennium, Jimi was going to Wisley as well; he says, 'That's always been a great inspiration for me, plantwise.' A particular memory for him was seeing the 150m (492ft) long Piet Oudolf borders at Wisley, which he describes as 'my first introduction to that sort of planting'. These two borders, which the Royal Horticultural Society commissioned Piet Oudolf to design, were completed in 2000. When they were first installed, they were quite revolutionary in Britain, as never before had grasses and a modern range of perennials been seen on such a scale.

Towards the end of his stewardship at Airfield, Jimi noted that more emphasis was being placed on the commercial aspect of the gardens. 'At that time I took a trip to England and after much soul-searching I decided that I had taken Airfield as far as I could take it. I felt it was now time; I was ready for a new project and to start a new chapter in my life,' he says. Later, in 2011, Airfield was closed to the public and following an ambitious, multi-million-euro redevelopment plan by garden designer Arabella Lennox-Boyd and landscape architect Dermot Foley, the garden was reopened. The ornamental gardens at Airfield are now in what Jimi describes as 'the expert hands' of head gardener Colm O'Driscoll.

STARTING OUT AT HUNTING BROOK

Once Jimi began to realize that things at Airfield were moving away from what he wanted the place to be, the decision to go it alone at Hunting Brook was reached fairly quickly. He remembers very clearly where he made it – the garden of Spetchley Park, in Worcestershire. Spetchley is a very special place, one of those big country gardens which has 'been in the family' for centuries, full of magnificent old trees, rather idiosyncratic plant collections, and perennials and bulbs which have naturalized with time; the thousands of *Lilium martagon* that grow in the lawns are an incredible sight. He says, 'I remember going into the little old-fashioned café there, having a ploughman's lunch and getting out my notebook and writing down what I would do if I left Airfield, how I would earn money. I just brainstormed – I loved Airfield, but I saw that everything was changing. Being at Spetchley was the turning point, as it made me see everything so much more clearly.'

Having been given 8 hectares (20 acres) of the family land, Jimi decided he would move there and establish a garden and teaching business. The land was more than he had dared hope for, although much of it was terrain that many might have fought shy of taking on. It is bisected by a very steep valley, with some quite dense tree planting. Jimi could remember playing in the area as a child, although as a farmer's son there was very little time for play. 'I called it Hunting Brook because we found a 1790 map with the name 'Hunting Brook' on the stream,' he explains. He imported a log cabin from Poland over the winter of 2001; he remembers that 'it came on the back of a truck and took seven weeks to build. It came with builders from Poland who lived in a caravan on site.'

By coincidence, or perhaps a case of 'great minds thinking alike', June opened a plant nursery around this time too. She had worked as a silversmith after leaving school and had her own business 'but always did some farming, then I became a full-time sheep farmer for most of my life, as well as having a family. I had always done a bit of gardening, but then I thought I should make a career change. None of the children was interested in taking over the farm, and I could not keep on working it as a farm myself. I had the idea of growing perennials from seed, running a nursery, so with Jimi's encouragement I did a basic RHS course at the Dublin School of Horticulture, and then worked at Mount Venus Nursery. I've learnt a lot from Jimi, although we garden in very different ways.'

The nursery grew a wide range of perennials from seed which was often bought from Jelitto, the German company which is the world's leading supplier. June also acquired seed from a number of plant hunters, such as Chris Chadwell, who worked with a seed collector in Kashmir – 'I still have a fantastic orange potentilla from him,' she says – and Ray Brown of Plant World nursery in Devon, who in 2002 went on a plant hunting trip to Sakhalin, in the Russian Far East; plants from that region grow well in Ireland. June's nursery business gradually segued into a garden business, attracting visitors for the same reasons as Jimi's. 'I'm much more selective in what I grow,' she says, 'but often he has something which I like, and then I'll get some cuttings off him.'

I was intrigued, as I'm sure many are, by the way the two siblings are doing such a similar thing at a similar time. 'Both gardens just evolved organically,' says Jimi. The gardens complement each other nicely as they are so different, and of course create a powerful synergy – those who drive out of Dublin to visit one inevitably end up at the other too.

Hunting Brook Gardens opened in May 2002, even though Jimi had only left Airfield the previous September. 'I had to plant beds really quickly,' he remembers, 'and I had to propagate an awful lot when I came here.' Many might feel diffident about opening a garden to the public after only working on it for a few months. Not Jimi; for some this might be sheer self-confidence, but in Jimi's case there is a different thinking about gardens at work. For him, a garden is never finished, but is to a greater or lesser extent always a work in progress. The development phase of a garden is a teaching opportunity, and education, at Airfield, had become a big part of what Jimi was about. Right from the beginning, he ran courses at Hunting Brook and the construction of the cabin was designed with teaching space in mind.

'It's not an easy way to make a living,' he says, 'but I don't have to answer to anyone. When I moved here I was designing gardens, doing consultancy, writing for a garden magazine, *The Irish Garden.*' That soon changed.

ABOVE
Jimi's sister June Blake working in her border of hot colours. "She has such a keen eye for colour and choice plants," says Jimi.

OPPOSITE
Jimi's home, centre for education and tea room under construction.

BILLY ENJOYING THE NEW BED

SOME HELPERS I WOULDN'T BE WITHOUT

FRED THE BEAUTIFUL COLLIE WHO WAS WITH ME GARDENING EVERY DAY FOR 17 YEARS

DORIS THE OLD ENGLISH SHEEP DOG, GREETS EVERYONE AS THEY ARRIVE AND WALKS THEM THROUGH THE GARDENS AND VALLEY

G-STAR

GROWING UP WITH PLANTS

APRIL 1979

This is me at Dublin Botanic Gardens after my first Holy Communion. My choice for a treat after the service was finished.

MY MUM ADORED HER GARDEN AT TINODE AND WAS ALWAYS GARDENING EVERY SPARE MOMENT SHE HAD

JUNE 1999

This is me standing with Fred at the main door of Tinode the year it was sold.

THE DAHLIA HAS ALWAYS BEEN A FAVOURITE OF MINE

AUGUST 2001

My house under construction after the timber arrived from Poland. The digger arrives to start the garden build.

DORIS LOOKS LIKE WE FELT

THE CABIN GOES UP

BUILDING HUNTING BROOK

SIGN IS UP AND WE'RE OPEN

AUGUST 2003

Early days at Hunting Brook. These large groups of perennials have gone through a few new looks since this was taken. The dahlia is 'David Howard'.

DOUBLE DAHLIAS IN FULL BLOOM

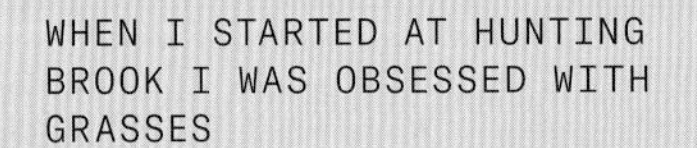

WHEN I STARTED AT HUNTING BROOK I WAS OBSESSED WITH GRASSES

THE FIRST SUMMER

SEPTEMBER 2007

Jimi and Ashley at Hunting Brook surrounded by a sea of lush foliage.

JULY 2017

My first cactus garden at Hunting Brook. I like to move them around and plant the cactus in different areas each summer.

CREATING A MEADOW FEEL IN FRED'S GARDEN

MIXING IT UP

MONTY DON, BILLY AND ME

JUNE 2018

Billy and I sitting in the woodland garden when Monty came to film BBC Gardener's World.

TEACHING COURSES AND LECTURING

Right from the start, running courses was part of the idea for Hunting Brook. Jimi says, 'I brought people in to teach courses on art, basket-making, weaving, yoga and meditation. I commissioned chefs from my favourite restaurants to hold classes, such as Dennis Cotter, from Paradiso in Cork, one of the best vegetarian restaurants in Ireland, or indeed the whole of Europe. For the gardening I had Carol Klein, Fergus Garrett and Chris Beardshaw.' Over the years, the garden has taken ever more of his time and his horticultural teaching commitment has very much taken over but other activities continue, particularly lecture tours and leading garden groups on trips abroad.

Jimi says that his courses attracted 'lots of people at a crossroads, women who had had their kids, people looking for a change in their lives'. In running them he was something of a pioneer, and again, very much the right man in the right place at the right time. We tend to take independently organized classes for granted now, but in Ireland in the early 2000s, they were a novelty. With growing income and leisure time, Irish society was just at that point when increasing numbers of people were able and willing to go on journeys of self-discovery, which might involve exploring food, well-being and gardening.

A key part of Jimi's educational programme has been the Plantsperson's Course, which he started in 2007. It consists of one day a month (minus December and January) and Jimi, always the innovator, says it will soon be available online. Jimi gives a presentation in the classroom, practical demonstrations and a class in the garden; one day of the course is also spent visiting other notable gardens. Copious plant notes accompany the teaching. It's a remarkable achievement in that the course offers the opportunity to learn about a very wide range of plants including their cultural requirements and their visual possibilities for garden-making. There is also a regular 'supper club' where Jimi gives an evening lecture or a tour of the garden and cooks a vegetarian meal.

OPPOSITE
Jimi teaching the Plantsperson's Course in the gardens at Hunting Brook.

I GET MORE OUT OF THIS THAN I EVER GIVE, WHETHER IT'S IDEAS OR SEEDS

The classes and the very social way in which Jimi engages with people have helped immensely in attracting volunteers. Voluntary labour is a key part of how many gardens function nowadays, the basic deal being that volunteers are doing something they enjoy and are learning from it too. Teacher Susan Dunne, whom I met one day when she was busy deadheading, is one of these. 'I love this garden,' she told me. 'About two or three years ago, Jimi put an appeal on Facebook for volunteers, so I came along to get my instructions and I have been coming ever since, every Thursday during the summer, when I'm not working.' She described how earlier in the year the volunteers were doing cuttings and potting on; now they were deadheading. She said, 'I get more out of this than I ever give, whether it's ideas or seeds. I have quite a large garden at home, so it's an inspiration being here. It's a wonderful spot this, full of magic and fairies, and I love everything about it, including Jimi.'

Orlaith Murphy came here to take the course and then learned that Jimi needed a part-time personal assistant. She works on the website, newsletter, tours, marketing and some of the admin. Jimi says of her, 'Orlaith is an amazing person to work with, particularly on the tours – she pulls it all together and is so good with people.' She is a keen gardener and has a particular interest in what she calls 'edimentals', plants that look good but are also edible. She says, 'I teach workshops myself on these plants and on foraging.

Charlie Clarke is a local man who at the time of writing has been a full-time volunteer for a year and half. 'He has been an incredible help, doing everything from general maintenance to planting and looking after the public when I am away,' Jimi says. Ciaran Farrelly was a student here back in 2011 and continued on working in the garden on many of the large projects. He's a passionate plantsman and helped Jimi start the tropical style of planting at Hunting Brook. Jimi describes Sam Hoey as 'an absolute plant nerd like myself, a local guy who came here on a college placement and has continued working in Hunting Brook on open days. He is cataloging the plant collection.' Sam has gone on to study at Wisley but continues to come back and help out. Oisín Orpen started as a volunteer when he was training in horticulture 'working in the garden a day a week, and since finishing his studies and setting up his own business he continues to work here once a week.' James Plunkett was graduate of Dundrum College and has being invaluable with his help maintaining Hunting Brook over the last number of years.

Jimi's lectures abroad started in 2010, with an invitation to the Perennial Plant Conference at Longwood Garden in Pennsylvania. The first American lecture is always a significant step in a garden media career. Most garden lecturing is on the East Coast or the Pacific North-west, though there are some exceptions, such as Spokane in eastern Washington State. We have both been speakers there and have experienced the hospitality of the Meyer sisters, who organize one of the most successful garden shows in the USA and use the funds generated to support an adventurous lecture progamme. The sisters are the most extraordinarily energetic, hospitable and genuine people. 'They know how to organize a good event,' Jimi remembers. 'It was like a wedding – all the food, all those cakes!'

OPPOSITE TOP TO BOTTOM
Exotic planting outside the class room with a mixture of foliage plants and dahlias. Jimi grows *Canna* 'Bird Of Paradise' on the left for its amazing foliage – the leaves are a slate blue-grey with purple veins and a purple edge.

Canna 'Bird of Paradise' at the front with *Dahlia* 'Lady Darlene', salvias and *Ensete ventricosum* 'Maurelii' at the back to create drama in this small space.

Orlaith Murphy works with Jimi on the garden tours programme, publicity and marketing.

ABOVE
Carol Klein came over to teach a course at Hunting Brook and is pictured here with Jimi and his mum.

TRAVEL TO SEE PLANTS IN THE WILD

In 2002, as soon as the builders of his house went back to Poland, Jimi set off for China for three weeks to join an expedition led by Seamus O'Brien, who was travelling in an official capacity for the National Botanic Gardens. Named the Glasnevin Central China Expedition 2002, the expedition started in Hubei province and ended in Szechuan, the aim being to follow in the footsteps of the Irish plant hunter Augustine Henry (1857–1930). Helen Dillon was with the ten-strong group too and the trip was filmed for RTE. In part they travelled down the Yangtze, which was at the time being flooded by the arguably disastrous Three Gorges Dam project which caused the local population to flee the flooding; the group passed people going down the river with all their possessions on the boats.

'We were taken to the top of Mount Omei,' Jimi recalls, 'and walked down, 250,000 steps, sideways –the Chinese have smaller feet than me. We made more than 450 seed collections. Some of them, such as *Aralia echinocaulis*, were endangered because of the flooding. We went back to the original area where plants of *Metasequoia glyptostroboides* were found in the 1940s, and got seed there too – they are now down in the Valley, and 15 foot (4.5m) high. There were some good hydrangeas, too.' The seed collecting was done with the full co-operation of the Chinese hosts, the Botanical Garden of Chonqing. 'We had to clean the seed in our rooms,' he says, 'and you know how bad gingko fruit smells, my room stank of dead sheep.' One of the places they went to was a farm in Hubei where there grows a 1000-year-old gingko, one of those trees whose origins are dubious, as whether planted or wild can be very difficult to ascertain; a seedling thrives at Hunting Brook.

Many people starting to plant gardens stay put. The demands of seedlings that need pricking out, young plants that need watering and weeds that need rooting out tend to hold people down to the patch of earth where they have chosen to create their dreams. Not Jimi. For him, travelling is a key part of garden-making. 'I need to find plants, to meet garden people, be inspired with fresh ideas – a lot of gardens get so stale, don't they?' he says. And a need for constant stimulation simply demands that he sets off on fresh journeys. One particular aspect of travelling that Jimi has built into his schedule is tour guiding; he

ABOVE
Jimi's love of dramatic foliage was brought to a new level after he spent a few winters in Indonesia.

has spent his life visiting gardens and this was a natural progression. So far he has concentrated on Ireland, Britain and Holland, with one trip to Sweden, but he is aiming at France and, rather more dramatically, Bosnia. He had reconnoitred the latter while we were working on this book. It would be planned as a botanical tour, in an area with remarkable biodiversity.

The British tours often involve a visit to Great Dixter and its Autumn Plant Fair. 'The perfect tour involves, ' he says, 'a mix of gardens and nurseries, no more than two destinations in a day. A lot of it is about buying plants – the last trip to Holland was amazing and people came home laden with so many plants that some had to be shipped as well. To do a tour I have to go on a research trip first. That's the fun part. The hard part is the logistics as we usually have as many as 30 people. That's why Orlaith comes along to help.'

Since Jimi started Hunting Brook, he has been over to England regularly, always visiting Great Dixter. 'I've really built up a relationship,' he says. 'There is no other garden that gives me such a strong feeling; for me it is the most healing garden of all. I love it in spring, when you can see the bones of it.' Recalling a recent visit, he goes on to say, 'I always learn something from Dixter. I learn a lot from the meadows and from their succession planting and working with pockets – those pockets are highly labour intensive.' It is the high-density succession planting that makes Dixter so special, the turnover of plants within a space that provides a constant stream of interest. 'There is some really good container planting going on there,' he says. Indeed, planting containers and moving them around is a key way that Fergus Garrett and his team try out new plant combinations before committing them to the soil and a greater degree of permanence. Jimi is also very interested in how they organize students, which is one of the most notable aspects of life at Great Dixter; many young people come every year to work there. Collaboration with students and interns is one way in which many gardens with limited funding are able to survive and continue to experiment. Indeed, there is plenty to learn here, as Jimi himself is starting to take on interns and students from all round the world.

DRINKING TEA WITH THE LOCALS IN RAJASTHAN

TRAVELLING

A KEY PART OF MY GARDEN-MAKING

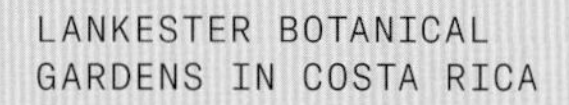

LANKESTER BOTANICAL GARDENS IN COSTA RICA

SEEING THE STRIPED LEAVES ON THE BANANA PLANT IN BALI INSPIRED ME TO START GROWING THEM

TOTALLY IN AWE OF THE
RHODODENDRON COLLECTION AT
BODNANT GARDENS IN WALES

ARTHUR'S PASS, NEW ZEALAND - ONE
OF THE MOST INSPIRATIONAL PLANT
COMMUNITIES I'VE EVER SEEN

THE LUNAR LANDSCAPE
OF NEW ZEALAND

A DAWN REDWOOD IN THE WILD

BABIES FROM THIS TREE ARE
NOW GROWING IN THE VALLEY
AT HUNTING BROOK.

Another garden that has been very influential to Jimi is Chanticleer, just outside Philadelphia. Established as a private garden in the latter part of the 20th century, it was left with a substantial endowment which has allowed its gifted leadership to develop one of the most exciting and innovative gardens in the Americas. Like Great Dixter, it is a high-intensity garden but in a very different way, being made up of several separate sub-gardens each with their own distinct character and management regime. Crucial to its success is the way that each area is managed by one individual who is allowed considerable autonomy; no one becomes too fixed in their ways, however, as every few years staff swap areas.

Jimi worked at Chanticleer during May 2009, a week in each of the sub-gardens. 'That is one of my favourite gardens, the one that inspires me most,' he says. 'It took me weeks to get it, as it's very complex. I was very impressed with how well the place is run – the independence given to the staff who are in charge of their areas really makes it. I love the terraces and also the exotic plants, which have inspired me a lot in what I've done here.' The sub-tropical planting around the house involves warm-climate species being planted out every year in a series of relatively small and defined spaces, making for an exuberant jungle-like effect but one which, as he notes, is dependent on 'the heat of American East Coast summer, although increasingly I feel we are able to do it here'.

'The woodlands are really good, too,' he says, talking of two areas that have a highly diverse shade-loving flora growing beneath trees. 'Lisa Roper is a great gardener – she was doing the Asian Woods area when I was there.' Another hit with Jimi is the cut-flower garden, which is best described as being like a cottage garden gone mad, with a wide variety of perennials and annuals in a state of organized chaos – 'giant cut flower garden' might be a better name for it.

Since he was at Airfield, Jimi has been particularly fond of visiting plant fairs – events where nurseries come to one site to sell plants. Sometimes attached to larger flower show events but very often plant sales and nothing else, these have become one of the most important ways in which small specialist nurseries can sell their wares and where keen gardeners can find rare and unusual plants. In Britain they really took off in the 1980s; I had a nursery myself at the time, and I found the sales organized by the Hardy Plant Society and the National Council for Plants and Gardens (now Plant Heritage) a wonderful way of both selling plants and making myself known. There are similar events now in Germany, the Netherlands (where Piet Oudolf was a pioneer organizer), Belgium and France – indeed France's Journée des Plantes, now at Chantilly but originally at the Domaine de Courson, can probably be described as Europe's finest. Ireland itself has a growing plant fair scene, with a Rare Plants Fair which has been held every spring since 2001 and moves around the country; there are also frequent fairs organized by the Irish Specialist Nursery Association.

BELOW
Jimi went to the Island of Flores in Indonesia with his friend Trevor and here he is kitted out to explore the caves.

OPPOSITE
Enjoying the scent of a magnolia in Tregrehan Gardens in Cornwall.

For the customer, plant fairs avoid the necessity to spend a lot of time and money driving around from one small nursery to another. For someone like Jimi, with a limited budget of time and money, they have been a particular boon. His current favourite is, he says, 'The plant fair at Tregrehan, the one that Tom Hudson puts on. That's an amazing event. All the small nurseries from that part of the world are there and I bring back enormous amounts of stuff.'

In analysing his life, he thinks he is 'an excitement addict – I have to do something that gets me really excited. I love to run, but I can't run in the same place twice . . . the excitement of new projects, new clothes, new food, that is what keeps me going.' And of course new planting projects. 'Right through my whole life, gardening is what I come back to,' Jimi says, leading on to an issue which is close to his heart – that of the therapeutic power of gardening and working with plants. 'In my twenties, once I left college I just partied hugely,' he recalls. 'I loved the Dublin social scene but there came a time in my life when I couldn't continue the excessive partying. So when the opportunity came to leave Dublin I was ready.' Seamus O'Brien remembers his parties as 'fantastic', but Jimi is insistent that 'I was always absolutely immersed in gardening, it helped keep me on track, something I could lose myself in, whatever else I was involved in . . . and then I suppose by my late twenties I got interested in yoga and food, and then feeling good, and feeling good began to become more important than wild weekends, but it was hard to move out of the city. Once I was out it took a few years down here to settle into country life away from the party scene.

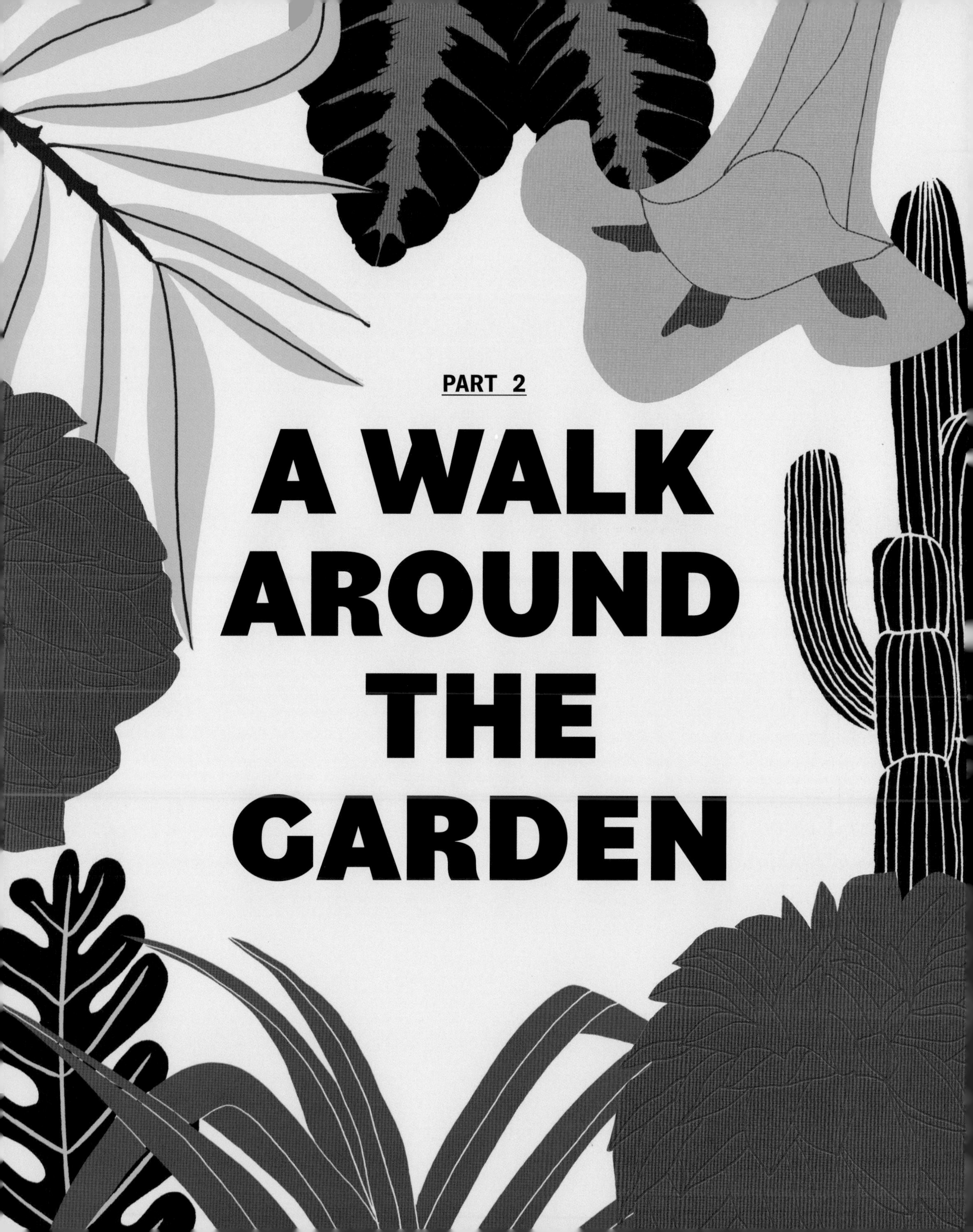

PART 2

A WALK AROUND THE GARDEN

FIRST IMPRESSIONS

Arriving at Hunting Brook, and walking up what feels like a drive but in time will almost certainly be more of a path, let's start with those very distinctive trees I mentioned before. They turn out to be *Aralia echinocaulis*, a Chinese member of a genus with which many of us are vaguely familiar. They look like a natural grove, as they are all different sizes and even display rather different shapes. Grown from seed collected in China on Jimi's trip in 2002, they are among the oldest plants in the garden and do much to create an atmosphere and sense of place. The grove may be moved around a bit, as Jimi has plans to pull in the borders from each side and transplant some of the smaller trees.

The walk up the drive is a slow one, as there is so much to see on each side. On my previous visit, the border on the right was newly planted, a sophisticated blend of apricots and soft oranges, although some very bright purple-pink *Geranium* 'Anne Thomson' run along the bottom, looking a bit out of place. Jimi explains to me later that these are coming out over the next winter as they are part of the last planting scheme here. On the left-hand bank a variety of herbaceous plants flourish; the overall impression here is of green, but with some yellow dahlias, bamboos and in particular, looking at it in August, big clumps of *Persicaria amplexicaulis,* that stalwart of late summer gardens on moister soils. I am later told that 'it's all coming out', so a big redesign is clearly going to happen there in the coming winter.

In the planting on the right, there are some cacti and succulents along the edge, looking rather alien. They are overwintered under cover, so it seems as if their role in life is to be endlessly moved around, creating a frisson of surprise wherever they appear. They reinforce the general impression of a garden that is bold, experimental and distinctly playful – and one where things move around rather more than they do in most gardens.

On reaching the top of the slope, it is hard to know where to look. The garden opens out and there is an exuberantly colourful border directly in front, but the eye is inevitably distracted by two very bright plastic chairs off to the right. Looking around, there is another pink distraction up by the house, a very large metal flower sculpture made by Jack Hart, an artist in Co. Wexford. 'I had a stand at the Bloom annual flower show one year, and bought it there,' Jimi explains.

It is now possible to look around and get an idea of where we are. There are trees most of the way around, so the impression is of being on the edge of a native woodland of oak and ash but one fronted by many unusual leaf shapes, an exotic frontage to the familiar including species of *Catalpa, Daphniphyllum, Toona, Rhus* and several bamboos. The trees, and Jimi's house, enclose the area on three sides, but on the fourth there is a view across to a typical Irish pastoral landscape with cattle grazing and trees on the other side. Since the road is set down in something of a dip it effectively becomes a ha-ha in an 18th-century landscape, so there is the illusion of Jimi's garden connecting directly with the field's grasses and rushes.

From August to October it is dahlias and salvias that dominate here, but the strong foliage of the surrounding woody plants is reflected in some of the planting too, in the shape of banana and canna species. Turning around to the left, a path leads to the front of Jimi's house, a log cabin with a large area of decking at the aralia-shaded front. It is a narrow path running along the top of the soon-to-be-redeveloped-border mentioned earlier, dominated by hardy perennials; I notice some happily spreading *Rodgersia* and some rather fine dark-stemmed *Eupatorium.* Here is another chance to look out, through the stems of the aralias (appreciating how moss-covered they are) to the landscape beyond.

PREVIOUS PAGE
Fred's Border plays silver foliage off a range of colours including the orange of *Calendula officinalis* 'Indian Prince'. *Artemisia stelleriana* 'Broughton Silver' and the rosettes of *Verbascum bombyciferum* line the path.

OPPOSITE
Cactus *Cleistocactus strausii* with *Calendula officinalis* 'Indian Prince', *Aeonium* 'Voodoo' (left) and *A. hierrense* (right). The creamy umbellifer just visible to the right of the frame is spontaneously-growing *Angelica sylvestris*.

BELOW
Allium 'Globemaster' and *Dahlia* 'Twyning's Revel' with the foliage of *Ensete ventricosum* 'Maurelii' and the spires of *Linaria* 'Peachy' in the background.

NEXT PAGE
The light canopy of *Aralia echinocaulis* allows perennials to flourish underneath including the grass *Cortaderia richardii*. The blue on the bank is *Geranium* 'Rozanne'. In the foreground are several spires of pale blue-purple *Veronicastrum virginicum* and the big yellow head of *Verbascum bombyciferum.*

AN OBSESSIVE COLLECTOR

PLANTING PHILOSOPHY

By now it is apparent that this is an example of what is sometimes called a 'plantsman's garden', but one with more attention to aesthetics than many that attract that label. The use of colour is bold, but careful and considered. There is also clearly a great interest in foliage, and it is particularly with respect to leaves that the more you look the more you see. Every nook and cranny seems to reveal new shapes and textures. Jimi would be the first to say that Hunting Brook aims to boldly foreground attractive new foliage plants, and to do so without being too worried about how hardy they are.

At base, Jimi Blake is an obsessive plant collector, but he is far more than that. His choices are tempered by a strong emphasis on planting design, with a particular feeling for both distinctive foliage and colour. After many changes at Hunting Brook, the planting style he has settled on for the open sunny areas of the garden is perhaps best thought of as a two-stage year: perennials and bulbs until July then a handover to mostly temporary later season plants, with a strong undertone of foliage. In Fred's Garden in particular, perennials and grasses provide much of the anchor plants into which he plants annuals and tender plants that have overwintered in the polytunnel. Jimi's affinity for mixing grasses and perennials goes back a long way and indeed in the early years of Hunting Brook he got newspaper write-ups of the 'the bright young thing with grasses' kind. His planting has evolved from those early days and he now prefers a more complex and visually richer look, but one which is inevitably much more labour intensive, filling in around the early-season perennials with annuals, temporary summer planting (led by dahlias and salvias) and exotic foliage. It's a planting style that involves a big May planting-out session and a big October dig-up. In that way it's something of a contemporary take on a Victorian way of doing things, a way which is rarely practised now, but with one notable exception – Great Dixter, where Fergus Garrett maintains this high-energy route to constantly changing, full and colourful borders.

RIGHT
Ashley's Garden combines strong foliage shapes such as *Ensete ventricosum* 'Maurelii' and *Echium pininiana* with colour, provided here by orange *Lilium tigrinum* 'Flore Pleno' and the dark red spikes of *Salvia confertiflora*.

INTERESTING SHAPES AND TEXTURES AROUND EVERY CORNER

ABOVE
Telanthophora grandifolia, one of the Central American foliage plants Jimi finds so useful for summer luxuriance.

RIGHT
A hybrid *Trichocereus* cactus with *Geum* 'Totally Tangerine', a very long-flowering variety of a familiar genus. Such juxtapositions are what makes Hunting Brook such an exciting and surprising garden.

ABOVE
Senecio cristobalensis is one of Jimi's favourite plants for its overall form and rather cuddly leaves; it needs winter protection.

SURPRISING COMBINATIONS THAT MIX THE EXOTIC WITH THE FAMILIAR ARE KEY TO JIMI'S STYLE

A WAY WITH WOODIES

So much for the borders, where change is a constant. What about the more permanent parts of the planting palette, the woody plants that form the long-term basis of any garden? As you walk around Hunting Brook you will see plenty of trees, for this is essentially the edge of a woodland, or indeed actual woodland. There are not many shrubs, however, at least not of the kind you normally find in gardens, especially informal country gardens like this. There are some witch hazels (*Hamamelis* spp.) and rhododendrons, but not much else. Those species that are woody and planted by Jimi are more likely to be very unfamiliar-looking, such as the aralias that tower over the first part of the garden you see as you walk up towards the house.

We should take a step back and think about the wider role of woody plants in gardens before looking at what Jimi is doing with them, which is actually very radical and probably very far-sighted. Indeed his use of woody plants is possibly his most revolutionary innovation of all.

So, why do we grow shrubs and small trees in gardens? The answers are obvious to most gardeners: for their flowers, to which the designer might add: for their structure and mass. The flowering season is actually quite restricted, from spring into early summer for nearly all garden shrubs, and subject to the great proviso that if you live on alkaline soil the whole wonderland of the heather and rhododendron family, the Ericaceae, is closed to you. And after flowering? Most garden shrubs are amorphous blobs, without good structure and with rather uninteresting leaves. Some look positively dreary and take up a lot of space doing so. 'I don't see the point of growing something so dull for a few weeks of flower,' says Jimi, 'I want year-round structure.'

Is this an unreasonable demand? No, not really. Historically, shrubs in gardens have tended to be evergreen and clipped – indeed the classic Italian or French garden was composed very largely, if not entirely, of such plants. So too is the Japanese garden, although with a very different aesthetic. I would suggest that Jimi is reinventing this concept but with a dramatically different range of species and using their natural form.

RIGHT
Sunrise over Hunting Brook. The distinctive New Zealand *Pseudopanax crassifolius* towers over *Monarda* 'Gardenview Scarlet', *Filipendula* purpurea and *Persicaria amplexicaulis* 'Orange Field'.

ABOVE
Billy being affectionate with Doris's snout just visible.

RIGHT
Wild *Angelica sylvestris* is allowed to seed into some of the planting below the *Aralia echinocaulis*.

OPPOSITE
Fatsia polycarpa in the Valley, one of many Araliaceae new to cultivation at Hunting Brook.

Historically, deciduous shrubs had a rather restricted role in gardens, largely banished to the outer and rather wilder reaches. Late 19th- and early 20th-century plant-collecting massively increased the range of shrub species and many of these found their way into the heart of gardens. How could they not, with such wonderful performances of flower? But the problem remained that for much of the year they were really quite uninteresting, as indeed were most of the evergreen shrubs if left unclipped. Designers would talk of how they created mass, their all-season shapes acting as walls and screens, framing views, guiding the eye and managing the whole experience of walking around the garden – all very true, but there was always that uncomfortable sense in the back of one's mind that this advice was really trying to make the best of a rather poor hand.

Philadelphus, Rhododendron, Syringa, Viburnum, Chaenomeles, Deutzia: however lovely they are in flower, they are dull blobs when not. In the smaller gardens that most of us have, this is not what we want. However, the last couple of decades have seen a growing range of woody species being introduced which have a far stronger sense of structure, overwhelmingly members of the ivy family, the Araliaceae. Many are from New Zealand or the Far East. Why so few of these were introduced back in the golden age of plant hunting at the end of the 19th and beginning of the 20th centuries is a bit of a mystery; indeed we only seem to have *Fatsia japonica* from this era, which gives a rough idea of the general Araliaceae look – big palmate leaves, dark, limited branching. Many are somewhat tender, although that never stopped the introduction and cultivation of camellias. There was a great love of the exotic at the time, and since these plants frequently look more tropical than they really are, the failure to introduce more of them seems all the more extraordinary.

Jimi's trip to New Zealand in early 2018 confirmed his love of evergreen structure plants. The country's flora is overwhelmingly evergreen and much of it has a strongly graphic quality; in addition there are few flowers. This aesthetic is clearly going to become much more important at Hunting Brook. Foliage is by no means always chunkily graphic, as many non-Araliaceae species have growth that is in fact is very fine, and often surprisingly dense with patterns quite unlike that of woody plants from elsewhere. Jimi mentions '*Plagianthus regius* – Derek Halpin, a friend of mine up in the mountains, much higher up than me, has had one for years. It forms a tight pillar; I suppose it is like a softer box or yew.'

Jimi has always had a fondness for pinnate leaves. He says 'I love that structure and the lightness of them, and if you prune them, you can get more light through them which you can't do with plants with big leaves . . . there's a delicacy, they're not blobby.' The most successful example of this use of pinnate leaves is the grove of *Aralia echinocaulis* dotted along the borders in the walk up to the house, which in fact creates one of the most immediate and lasting impressions of the garden. Shrubby aralias are rare among hardy woody plants in that they are single stem (only *Hydrangea aspera* has a similar architecture and that soon spoils it by suckering) with large doubly-pinnate leaves in a clump at the top. The overall shape is rather like that of a palm – strong visual impact but very little shade cast. Eventually the plants sucker but not so much that the effect is lost. Large panicles of creamy flowers, followed by dark berries, add to the unusual effect.

ABOVE LEFT
The new growth of *Aralia echinocaulis*.

ABOVE RIGHT
From left to right, the leaf undersides of *Rubus lineatus*, *R. calophyllus* and *R. henryi* var. *bambusarum*, all relatives of the familiar blackberry and raspberry.

The narrow upright thrust of the aralias is echoed in certain species of the New Zealand *Pseudopanax*, such as *P. crassifolius,* which have an approximately similar form, but eventually branching at height. With their bizarre minimal leaves, thought to be an adaptation to reduce predation by moas (extinct giant flightless birds) these are plants with an aesthetic that is perhaps best described as 'challenging'. Like the aralias, they hardly take up any space in a horizontal dimension, at least until they are pretty much mature, and they have the additional quality of being evergreen. All these plants can be used to create vertical interest but with minimal competition being afforded to perennials or whatever else is growing beneath them. While in New Zealand, Jimi visited a garden on South Island; I arrived a month after him, to find that the owner had been most impressed by his suggestion that she use the locally native *P. crassifolius* to dot plant into a perennial border to give it winter and vertical impact. 'I'd have never thought of that,' she said.

Beyond these exceptional plants, most of the woody plants that Jimi is using are more conventional in that they have bulk as well as height. The majority are evergreen, and most, certainly all the Araliaceae, have a sense of architecture that is sufficiently strongly defined to avoid them becoming blobs. Looking through the list of favourite shrubs, there are quite a few with pinnate foliage –deciduous species such as *Ailanthus* as well as evergreen, for example *Mahonia*, which continues to be a source of striking high-quality foliage for a variety of habitats. There are those with palmate foliage (indeed most Araliaceae are palmate) such as an *Aesculus*, and on a less grand scale a good number of *Rubus*. The latter is a useful genus, as they are vigorous (sometimes overly so) and offer a great range of form and foliage quality, the elegantly pleated *R. lineatus* being a particular favourite.

Then there those with leaves of impressive dimensions, such as the large-leaved rhododendrons which Jimi is aiming to distribute down through the Valley. They are joined by several cultivars of *Daphniphyllum*, which look so much like evergreen rhododendrons in leaf but in fact are completely unrelated. *Hydrangea* is a genus appreciated for leaf too, especially the *H. aspera* types with their very large, dark, stiffly hairy foliage. Most hardy evergreens have notably smaller leaves and these make up a goodly proportion of Jimi's shrub flora: species *Camellia, Osmanthus heterophyllus, Trochodendron aralioides* and several species of *Viburnum*. Jimi believes camellias to be underrated for their foliage while other evergreen shrubs are often not looked at critically enough; he feels that the Chinese form of *Trochodendron aralioides* is much better than the Japanese form more usually available.

The limited number of deciduous shrubs that get a look in are chosen for their impressive or highly distinctive leaves: various poplar cultivars and a few willows, such as *Salix moupinense* and *S. magnifica*. However, there are very few of the deciduous 'blob' shrubs which make up the bulk of the conventional garden shrub flora, just a few *Berberis* (with superior foliage) and a scattering of others.

Bamboos may not be technically woody plants, but they function as such in their permanence in the landscape. They are also evergreen and have good structure, so they can be regarded as honorary evergreen shrubs. Several make a good impact here. Jimi likes them because they are light and have nice verticals, to which I would add that they have a grace and an ability to move which contributes an elegant touch wherever they are planted, valuable as a contrast to the rather immobile dark forms of many evergreens.

BELOW
Two exceptional foliage shrubs, *Trochodendron aralioides* Chinese form (left) and *Salix moupinense* (right).

JIMI HAS DITCHED A LARGE CHUNK OF THE CONVENTIONAL GARDEN FLORA IN FAVOUR OF STRONGLY GRAPHIC PLANTS

So Jimi has ditched a large chunk of the conventional garden flora and is urging us to try a different range of species. They fit a similar design niche to the clipped box, Portuguese laurel and yew of old: evergreen and structural, providing continuity 365 days of the year. The style is of course totally different in that it has nothing to do with straight lines or geometry, but the underlying role in the visual composition of planting design is similar.

Often slow to propagate, and with hardiness not yet fully evaluated, the 'new evergreen' flora is still largely unfamiliar and untried. It is part of a long-term trend, however, which in Ireland and Britain can be mapped by following the slow rise of the New Zealand evergreen. Hebes grew in popularity as garden plants during the 20th century, arriving in the standard palette of landscape plants by the last decade. Micropropagation then allowed cultivars of *Phormium tenax* to be used on an increasing scale from around 2000 onwards. Milder winters encouraged increasing use of *Pittosporum* cultivars, and breakthroughs in propagation enabled *Pseudopanax* and more Araliaceae to be used. Jimi's demonstration of the great potential of these strongly graphic plants could be the beginning of a major new trend.

Finally, we need to briefly consider some short-lived shrubs. Many members of the mallow family are woody and flower for months, but they fail to regenerate themselves from the base in the way that most shrubs do and after a number of years deteriorate and die. They will tolerate cutting back by two-thirds in the spring. A number of cultivars and hybrids of *Sphaeralcea* work well in Jimi's late-season planting schemes, including *S.* 'Childerley', with soft apricot-orange malva flowers for most of the summer and grey-green leaves, and 'Newleaze Coral' with soft green foliage and coral flowers; the latter is very drought-tolerant.

OPPOSITE
A *Chusquea* sp. guards the entrance to the Woodland Garden. *Chusquea gigantea* (inset) grows down in the Valley.

ABOVE
Linaria 'Peachy' with *Sphaeralcea* 'Newleaze Coral' – two relatively new varieties which show how Jimi likes to seek out cultivars with distinct and subtle colours.

1

3

WOODY PLANTS

WITH STRIKING LEAVES

1. *Aralia echinocaulis*
2. *Decaisnea fargesii*
3. *Olearia lacunosa*
4. *Toxicodendron vernicifluum* (note: potentially dangerous skin irritant)
5. *Pseudopanax crassifolia*
6. *Zanthoxylum ailanthoides*
7. *Frangula alnus* 'Fine Line'
8. *Pseudopanax lessonii* 'Touthera'
9. *Pistacia chinensis*

2

THE "NEW EVERGREEN" FLORA IS LARGELY UNFAMILIAR AND UNTRIED BUT PART OF A LONG-TERM TREND

DORIS ENJOYING THE SHADE

4

JIMI LIKES THE STRUCTURE AND DELICACY OF PINNATE LEAVES AND FINDS THEY RESPOND WELL TO BEING PRUNED TO ALLOW MORE LIGHT THROUGH.

5

...MORE WOODY PLANTS

6

SLENDER LEAVES CREATE VERTICAL INTEREST WITHOUT COMPETING AGAINST THE PERENNIALS THAT GROW BENEATH THEM

7

8

1

FAVOURITE DECIDUOUS SHRUBS

AT HUNTING BROOK

2

3

JIMI CHOOSES SHRUBS WITH A STRONGLY DEFINED ARCHITECTURE THAT ARE LESS LIKELY TO BECOME BLOBS

4

1. *Lindera triloba*
2. *Hydrangea aspera* from Gong Shan
3. *Oplopanax horridus*
4. *Kalopanax septemlobus* f. *maximowiczii*
5. *Plagianthus regius*
6. *Styrax formosanus* var. *formosanus*
7. *Tetracentron sinense* var. *himalense*

PLANTS ARE CHOSEN FOR THEIR IMPRESSIVE OR HIGHLY DISTINCTIVE LEAVES

FUNKY FOLIAGE PLANTS

FOR THE 21ST CENTURY

1. *Aralia vietnamensis*
2. *Oreopanax echinops*
3. *Schefflera kornasii*
4. *Pinus montezumae* 'Sheffield Park'
5. *Magnolia macrophylla* subsp. *ashei* x *macrophylla* subsp. *dealbata*
6. *Oreopanax incisus*
7. *Fatsia polyphylla* Giant Form
8. *Pseudopanax lessonii* 'Moa's Toes'
9. *Zanthoxylum alatum*

SPINY, ROUNDED OR PALMATE LEAVES ALL HAVE A PLACE AT HUNTING BROOK

WITH COLOUR AND TEXTURE

EXTENDING THE SUMMER

Annuals form a major part of Hunting Brook's flora and a select few varieties are used to play a crucial role. Midsummer is the time to note them, as this is when the spring sowings of hardy annuals really begin to get into their stride in Ashley's Garden and Fred's Garden. *Orlaya grandiflora* is a good example of what could be described as one of the 'new annuals' – plants promoted for and by flower arrangers as well as gardeners. Jimi likes its 'lovely, pure white flowers in big, flat-topped clusters, which look like lace-cap hydrangeas' and 'its extremely long flowering period . . . it will often flower until the first frosts, from either an autumn or spring sowing'. One of the great uses of annuals in plantings that are dominated by longer-lived plants is potentially that of rhythm plants, as the slight habit of growth that many of them have is ideal for distributing a colour, a shape or a texture as a theme through a planting.

'I've gone through phases with perennials,' says Jimi, when I ask him about what I identify as gaps in his plant palette – which is perhaps another way of saying 'Why aren't you growing more of the plants that everyone else thinks are really cool?' or even 'Why don't you grow the ones I like?' There are a few monardas in Ashley's Garden, so this seems like a good place to start; Jimi really rates 'On Parade', and indeed a few others: 'Balance', 'Purple Roster', 'Scorpion', 'Jacob Klein' and 'Gardenview Scarlet'. But then he states, 'I've had lots of monardas, but not many left now – and phloxes, I had many of those over the years too, but I do find it hard to use plants that don't have much going on before flowering or afterwards. I love the intense colours of phlox, but they are a blob.'

Jimi is a huge fan of late-season perennials but he's picky and doesn't use them all. 'As for asters, they may be lovely but if they take up space all summer I don't want them. For things to be in these beds they have to look good growing, look good flowering and have something to offer afterwards.' To help achieve this he has even started to keep asters in pots, to drop them into a border while they are in flower. He is also prepared to dig up perennials, at least fibrous-rooted ones, at any time in the growing season in order to make space or find a better place for them. It has to be said that this is unlikely to be a successful practice in drier climates.

RIGHT
Rudbeckia hirta 'Prairie Glow' and self-sown purple *Agastache foeniculum* lean out over a path in Fred's Garden. A dahlia seedling grows opposite.

Any perennials at Hunting Brook from now on are here for a jolly good reason. Jimi feels that *Foeniculum vulgare* 'Purpureum', the bronze variety of the herb fennel, adds a lovely softness to a planting design, but when the plants grow taller and before they set seed he cuts them down to avoid vast numbers of irritatingly deep-rooted little seedlings. Some persicarias earn their keep: *P. amplexicaulis* 'Fat Domino', a bulky perennial for sun or semi-shade that is an improved form of a widely grown species, is one such. 'Rosea', a pink form, and 'Blackfield', a relatively new introduction with blood-red spires, are described by Jimi as plants that are 'invaluable for their long season of flower in late summer and into autumn'. *Eryngium eburneum* gets garden space for rosettes of finely toothed leaves with 1.5–1.8m (5–6ft) stems and greenish/white flowers, and some of the very best seedheads for winter. It grows at Hunting Brook in practically no soil and a very dry site. *Calamintha nepeta* 'Blue Cloud' is, Jimi says, 'a useful plant for the edge of borders, with scented leaves and soft blue flowers'. *Eupatorium maculatum* Atropurpureum Group earns a place as 'a classic tall late-flowering perennial for full sun and soil that does not dry out . . . one of the best plants for butterflies and it's a good cut flower'. *Artemisia* 'Powis Castle' is valued for its low mounds of silver fine-cut foliage, one of the few silvers Jimi uses. Like many others, he also appreciates *Salvia uliginosa* for its light, true blue flowers borne on rangy stems – awkward in the conventional border, but in denser plantings its stems can be supported by other plants.

OPPOSITE
A dahlia raised from Jimi's own seed and magenta *Cosmos bipinnatus* 'Dazzler'.

ABOVE
Extreme contrast: *Aeonium* 'Voodoo' with silver *Artemisia* 'Powys Castle'.

1

4

2

CLEAN, WHITE DAISIES
IN OCTOBER & NOVEMBER

LATE-FLOWERING PLANTS

TO EXTEND THE SEASON

3

CHALKY WHITE DAISIES
THROUGH TO AUTUMN

5

6

1. *Salvia stolonifera*
2. *Epimedium membranaceum*
3. *Leucanthemella serotina*
4. *Nipponanthemum nipponicum*
5. *Chrysanthemum* 'Ruby Mound'
6. *Eurybia x herveyi* syn. *Aster macrophyllus* 'Twilight'
7. *Hedychium forrestii*
8. *Geranium* 'Anne Thomson'
9. *Geranium wallichianum* 'Havana Blues'
10. *Aconitum carmichaelii* 'Royal Flush'

DAHLIAS GROWN FROM SEED
ARE INCREDIBLY GOOD VALUE

1. *Salvia* 'Amistad' – ***5 months***
2. *Dahlia* varieties – ***5 months***
3. *Linaria* 'Peachy' – ***5 months***
4. *Astrantia major* 'Bo-Ann' – ***7 months***
5. *Anisodontea* 'El Royo' – ***3 years!***
6. *Geum* 'Totally Tangerine' – ***7 months***
7. *Geranium* 'Anne Thomson' – ***6 months***
8. *Geranium wallichianum* 'Crystal Lake' – ***4 months***
9. *Geranium wallichianum* 'Havana Blues' – ***4 months***
10. *Persicaria amplexicaule* 'Orange Field' – ***5 months***
11. *Salvia* 'Flower Child' – ***5 months***

4

TOP PLANTS FOR LONG FLOWERING

AT HUNTING BROOK

IN FLOWER FOR 3 YEARS!

5

6

SALVIAS

LATE SEASON AND INTENSE COLOURS

Salvias are the single most important plant group for the late season at Hunting Brook – although, to be precise, it is the *Calosphace* subgenus of the Americas that Jimi, like many others, has fallen in love with; the *Salvia* genus is a huge one (1000 species globally) and clearly its existence is not going to be supported by the genetic research which is currently revolutionizing our knowledge of plant evolution and relationships.

Like orchids and a number of other groups, salvias have benefited enormously from the very complex corrugation of hills and valleys which geological history has thrust at the western strip of the Americas. From California down to northern Chile, populations have continually become isolated, mainly by climate, as the western face of a mountain range (a *cordillera*) is usually wet and the eastern very dry. Isolation allows genetically flexible plant groups to continually evolve.

We can have some understanding of why there is so much variation among salvia species, but there is no satisfactory explanation of their extraordinary colour range. Is there any other genus which includes such intense true blues as well as the deepest scarlets and crimsons and eye-searing magentas? Only good yellows seem to be missing. In all salvias the bracts often play an important part in the appearance of the plants too, sometimes being strongly coloured or having a furry texture.

Salvia Summit III was an event held in 2016 in Berkeley, California, with speakers from three continents. 'The salvia thing for me really took off after that Salvia Summit,' says Jimi. 'All the salvia nerds from around the world were there. I met Rolando Uria, an agronomy professor at the University of Buenos Aires – he's introduced a lot of really good species into cultivation, and he does an amazing list of salvias. I also get them from Dyson's in the UK. I've tried 256 species and cultivars, but I'm bringing that down as what I am interested in are ones that will combine well with perennials.'

Jimi adds that there are 'a lot of boring salvias with little blue flowers' as well as a good many which do not get enough heat to flower early enough in the cooler summers of north-west Europe. A key factor about salvias which he has recognized is that the species have

RIGHT
Salvia involucrata 'Mulberry Jam' is a recently developed cultivar of a Brazilian species which needs minimal winter protection.

different architectures which can have quite an impact on how they are used. 'Good ones for borders need to have light foliage, scattered up the stem so that they can combine well with other plants. Some of the new cultivars like 'Love and Wishes' – in fact all the 'Wishes' series coming out of Australia – have their leaves low down so they mustn't be covered by other plants, but they are good for the front of beds or containers.'

Jimi describes salvias as the easiest plant on the planet to propagate, even the old wood roots – which is just as well, as hardiness is always an issue. USDA zones of 6, and possibly 5, are the lowest that the hardiest of these species will tolerate. The majority are far less hardy, with many tolerating only very light overnight frost. 'They like plenty of light, they do not like to be cluttered,' says Jimi. My own observations in Mexico suggest that in nature their growth is often extremely sparse, with lanky stems bearing a few scruffy leaves and flowers at the top. Conditions in cultivation should guarantee good all-round light and gardeners should be prepared to do a lot of pruning, as many species need plenty of encouragement to stay compact. Jimi also recommends 'not giving them a rich soil, just well-drained good garden soil with no added manure or fertilizer as it deters them from flowering. They also need constant dead heading.'

'The involucratas do well here – they are a good group, so long-flowering,' enthuses Jimi, 'ones like *Salvia involucrata* 'Bethellii'. *S. guaranitica* 'Super Trouper' is another really good one, with a long series of blue flowers. *S. fulgens* was new to me last year and was the longest-flowering salvia, flowering from June till winter – it was the winner of my salvia trial! It grows up to 11,000ft (3350m) in Central Mexico, so I left it outside last winter and it had no problems with minus 5C (–23°F).'

ABOVE
Salvia involucrata 'Boutin' is an old-established variety of this very reliable species.

RIGHT
Salvia 'Kisses and Wishes'; the Wishes series of salvias has been bred by nurserywoman Sarah Knott in southern England.

SALVIAS THAT WITHSTAND COMPETITION FROM NEIGHBOURING PLANTS IN BORDERS

1. *Salvia* 'Amistad'
2. *Salvia atrocyanea*
3. *Salvia confertiflora*
4. *Salvia curviflora*
5. *Salvia dombeyi*
6. *Salvia* 'Envy'
7. *Salvia fulgens*
8. *Salvia guaranitica* 'Super Trouper'
9. *Salvia involucrata* 'Bethellii'
10. *Salvia involucrata* 'Boutin'
11. *Salvia involucrata* 'Mulberry Jam'
12. *Salvia patens* 'Guanajuato'
13. *Salvia* 'Phyllis Fancy'
14. *Salvia stachydifolia* (various colour forms)
15. *Salvia stolonifera*
16. *Salvia* 'Waverly'

ABOVE
Salvia involucrata 'Bethellii'. Another cultivar of this tall (1.2m/4ft) and relatively hardy species of salvia.

OPPOSITE
Salvia curviflora, a semi-shrubby species to 1.5m (5ft), recently introduced from Mexico.

ABOVE
Salvia fulgens grows to 1.5m (5ft), and flowers from mid-July until winter.

ABOVE
S. patens 'Guanajuato' is a selection from the best blue species, with a herbaceous rather than shrubby growth habit.

LEFT
Salvia microphylla 'Cerro Potosí' is a form of a relatively hardy species that forms a shrub to around a metre (3ft) in height.

LEFT
Salvia 'Amistad' has long been a very popular hybrid, much appreciated for its long growing season, bred by a grower in Buenos Aires.

ABOVE
A *Salvia patens* seedling from Jimi's saved seed.

DAHLIAS

LONG-FLOWERING AND VARIED

Along with salvias, dahlias are the great, late-season adornment of the borders at Hunting Brook. They vary enormously in the colour, size and shape of their flowers; indeed their flower forms are some of the most extravagant of all cultivated plants. For Jimi, 'the big thing is that they are long-flowering – if you pot them up early under cover, you can get them in flower in June, and they'll go on for months then.' In particular he chooses varieties 'bred for the cut-flower trade, as they have long stems, flowers held well above the foliage, they don't need staking, and each flower lasts a long time. In many varieties the flowers don't keep.' Dahlias are genetically so varied and so easy to breed that there is a constant output of new cultivars, and more than with anything else he grows, Jimi is always onto the best new thing. He is ruthless in dropping a cultivar if a better one comes along, and with dahlias especially, there always will be something better on the horizon.

Although there are plenty of showy doubles, mostly in Ashley's Garden, Jimi is now concentrating largely on singles. Unlike the doubles, singles are easily pollinated by bees and so produce plenty of viable seed, some of which Jimi grows on every year. In the past, anyone growing dahlias would have used only cultivars, but Jimi has joined an emerging trend in horticulture in exploiting the genetic diversity of growing unnamed seedlings. 'I got a mix of seeds from Keith Hammett, the New Zealand breeder,' he told me, 'and seed from the Botanical Gardens in Paris and from Mexico.' So alongside the predictability and consistency of some favoured cultivars are plenty of unnamed seedlings, all subtly different to each other. Among these are some with a very slight doubling, a characteristic that came with the seed from New Zealand – enough to look a little different, but not enough to hinder the ability of a bee to get to the nectar.

TOP
Collecting dahlia seed. The seed germinates rapidly when sown in spring, under cover at 20°C (68°F).

MIDDLE
Dahlia 'Lady Darlene' has huge flowers (20cm/8in) on a metre-high plant.

BOTTOM
Seedlings of Jimi's own dahlias. These are always singles as doubles have to be hand-pollinated.

JIMI HAS JOINED AN EMERGING TREND IN EXPLOITING THE GENETIC DIVERSITY OF GROWING UNNAMED SEEDLINGS

LEFT
One of Jimi's many dahlias grown from seed he has collected.

ABOVE
Dahlia 'Karma Bon Bini'.

ABOVE AND OPPOSITE
Two of Jimi's dahlia seedlings. Growing from seed results in a range of slightly different colours, creating an attractive effect when they are all planted out.

OPPOSITE
Dahlia 'Bright Eyes', also pictured above with a rear view of *Cosmos* 'Tango'.

LEFT
Dahlia 'Honka Surprise'.

ASHLEY'S GARDEN

Occupying the site of the original car park, Ashley's Garden now feels like the natural heart of Hunting Brook, with the house to one side and the woods behind. Ashley was Jimi's partner during the early stages of Hunting Brook and he sadly passed away. Although they separated they remained close and this colourful and vibrant border is named in his memory. 'He loved the garden so much,' Jimi says, 'he did a bit of work on it too, particularly steps and paths down the woods.' It is a February to November border, either looking its best or being nothing at all. 'I don't really go for the winter look here,' explains Jimi. 'I don't keep seedheads and I want to see the first signs of spring as soon as they appear.'

In late summer the eye is drawn first to the vibrant colours of dahlias, salvias, lobelias and big leaf shapes, chiefly cannas and bananas. When you look higher, though, you discover a magnificent backdrop of tree foliage. The background is chiefly composed of the ashes and oaks of a small area of woodland, but in front there is a middle ground of pinnate-leaved *Rhus verniciflua* and *Toona sinensis* and a magnificent bamboo, a *Chusquea* species. A couple of the dark spiky New Zealand *Pseudopanax crassifolius* are growing to maturity, with a steadily widening clump of foliage held a good 3m (10ft) above the ground.

However, you notice this scene-setting a bit later, for it is inevitably the contents of the border here which grab the attention for some while. 'It's basically a combination of perennials with big leaves and salvias and dahlias,' says Jimi. 'The idea is lots of colour for the late summer going into autumn.' He refines this, saying that he would rather have plants with a flowering season which extends well, or which follow on with good seedheads or foliage, rather than those which simply flower late. The latter can be unreliable, since one outcome of Ireland's cool summers is that some plants simply do not get enough heat to stimulate flowering until it is almost too late. On my first professional trip to Ireland I remember someone bemoaning the fact that in some summers even goldenrod did not flower.

PREVIOUS PAGE
A vibrant clash of reds and pinks – the shaggy red flowers of *Monarda* 'Gardenview Scarlet' with orange-red spikes of *Persicaria amplexicaulis* 'Orange Field' and puffy clouds of pink *Filipendula purpurea* 'Elegans'.

ABOVE
Ashley's Garden in the winter of 2002, with Fred.

RIGHT
The umbellifer is biennial *Angelica sylvestris*, descended from the dark-leaved form 'Vicar's Mead', with *Monarda* 'Gardenview Scarlet'.

Looking more closely at Ashley's Garden, it becomes clear that there is something of a paradox. Among the flower colours are a lot of pinks and oranges, two colours which conventional wisdom says should not be put together. Yet here they work. 'It's all very green,' I overhear a garden visitor say, and this indeed is part of the reason. With the exception of a very few double dahlias, every flower here has the shape and size of the natural species, and so the spots of colour are small and spattered across a mass of green leaves, which separate the colours and so stop any clashing effects. The garden design advice of not putting oranges (or yellows) next to pinks is based on an assumption of using many hybrids with artificially large flowers or flowerheads. There are in fact few yellows here – indeed there are not many in the garden as a whole, nor blues. The relative absence of these two very common flower colours narrows the spectrum down to a range of more interesting shades, ones which are not so common and which display a greater range of subtlety and variation.

The foliage matrix for all this colour is basically a mass of mid-green medium-sized leaves, with not many of the smaller leaves commonly found on summer-flowering perennials, a very few dark-toned ones and no variegation. However, dotted among them are some much larger and more dramatic leaves – the cannas and bananas most of us recognize but also quite a selection of other large-leaved plants, most of which look distinctly unfamiliar.

Come spring, however, and the border is all pulmonarias, hellebores, primulas (chiefly *Primula elatior* and 'June Blake') and bulbs, lots of bulbs: white daffodils to start off with, then tulips. 'I put in around 2000 tulips every year,' says Jimi, 'then alliums – you can save time by popping them both into the same hole.' The perennial components stay put, and indeed in late summer it is not too difficult to spot flashes of silver from pulmonaria leaves among the foliage of the taller late-performing plants. Plants of these three genera are permanent components of the bed, and are able to tolerate some shading from those growing later. Geums make quite a splash too, but being less able to tolerate shade are more likely to be on the outside. Cow parsley (*Anthriscus sylvestris*) is a feature in May. Jimi and I agree that this makes a fine element in planting at this time, the delicate tracery of its cream heads weaving between other plants and helping to provide a visual buffer between stronger colours.

LEFT
Musa sikkimensis 'Bengal Tiger' with *Astilbe chinensis* var. *tacquetii* 'Purpurlanze', *Lobelia* x *speciosa* 'Hadspen Purple' and *Dahlia* 'Karma Bon Bini'.

ABOVE
Musa basjoo pokes out above a cloud of flowers including *Thalictrum* 'Splendide'. Grass *Chionochloa rubra* emerges from the left.

May is changeover time. This border now sees a mass of new planting, for the majority of its later-season interest is the result of those plants that are here only temporarily. At the same time as the dahlias, salvias, cannas and so on going in, this year saw *Lychnis coronaria* 'Gardeners' World' and *L.*'Hill Grounds' (double dark red and bright pink, respectively) planted out now as well, to give colour for May and June; these were removed later after they had finished flowering, to be kept in pots for use again next year. 'The cow parsley gets cut back now to stop it seeding,' Jimi explains.

From August onwards it is the hot-climate exuberance of dahlias and salvias which dominates, with a wide range of warm tones – reds to yellows, and very few blues or purples. A closer look, however, reveals a lot more going on. This year it was obvious that Jimi had used *Cosmos sulphureus* 'Tango' as his continuity plant – up to 1.2m (4ft), rangy growth with neat divided leaves and small orange flowers. Every year, a different plant is used for this purpose; the previous year it was *Cosmos bipinnatus* 'Dazzler'. Several plants of *Actaea cordifolia* 'Blickfang' make a strong impact, not for their muted green and cream colours but for the extraordinarily vertical thrust of their flower spikes, which look as if they are determined to shoot off leaving a trial of sparks behind them, 'This is one of the best plants I got from Hessenhof, a nursery in Holland,' says Jimi, of what must be one of the best rhythm-marking perennials of all, each clump a visual drumbeat.

The late-season exuberance is given particular force by the wide variety of dahlias, mostly single and mostly unnamed seedlings, and salvias. 'The salvias here are all the ones that don't mind big plants around them,' says Jimi, 'and they have a relatively airy quality, such as 'Mulberry Jam', 'Amistad', *S. fulgens* and *S. curviflora*.'

An airy quality is an important aspect of most of what is growing here; it reduces competition and lets enough light through for the permanent perennials. On closer examination these are very much in evidence, although inevitably playing something of a background role now. There are a few of the broad fluffy spikes of *Astilbe chinensis* var. *taquetii* 'Purpurlanze'; normally its deep magenta heads scream for attention, but now they are more part of a background rhythm, along with pink-purple *Monarda* 'On Parade'. *Astrantia major* 'Bo-Ann' plants (pale pink and sterile, and therefore non-seeding) flower off and on through the summer too, while *Angelica sylvestris* makes a useful gap-filler. Various cultivars of *Sanguisorba* play an important role too, both for their very attractive early-season foliage and their great open heads of bobble flowerheads. These often flop about, which seems to bother Jimi less than it does most gardeners; if they become a nuisance they are cut down.

PREVIOUS PAGE
Allium 'Purple Sensation' and *A.* 'Mount Everest' flowering in May with *Musa basjoo* displaying its first leaves of the season.

LEFT
Dark red *Tulipa* 'Lasting Love' and orange 'Ballerina' with *Narcissus* 'Silver Chimes' in early May.

ABOVE
Salvia fulgens (left), *Dahlia* 'Ludwig Helfert' (top right) and a *Sanguisorba officinalis* seedling with *Thalictrum* 'Splendide' (bottom right).

Foliage is crucial to the look of Ashley's garden. Big chunky graphic shapes are dotted through, yet they are probably less than 5 per cent of the whole. What they add is a sense of punctuation. Several varieties of banana are used, including the red-leaved banana relative *Ensete ventricosum* 'Maurelii' and *Musa sikkimensis* 'Bengal Tiger'. Cannas have similar foliage, especially *C.* 'Musifolia', while *C.* 'Taney', with elegantly spaced and slightly grey leaves, is repeated several times. Slightly less obvious are several rarely seen large-leaved trees, many of them of Central American cloud forest origin – plants Jimi has been given by another Irish gardener. Once noticed however, these majestic plants do really stand out. *Telanthophora grandifolia*, the giant groundsel tree from Costa Rica with purple leaf stems, *Cyphomandra betacea*, the tree tomato, and *Oreopanax echinops*, a Central American member of the Araliaceae, are just three examples.

Senecio cristobalensis is a favourite with Jimi. A leggy Mexican member of the daisy family with slightly maple-like leaves coated in unusual purple hairs, it has many of the qualities that Jimi is good at spotting in foliage plants – there is nothing else quite like it, but it is not so unusual that it screams for attention. It is half-hardy and has to be dug up and moved into the polytunnel during the winter.

Among the hardy and therefore permanent foliage elements are a *Pterocarya stenoptera* 'Fern Leaf', potentially a large tree but coppiced annually at just below shoulder height, and eight plants of the grass *Chionochloa rubra*. This is the perfect tussock grass, its brown leaves arching elegantly up and over year round, reaching nearly 1m (3¼ft) across. Jimi brought the chionochloas from Airfield, memories of a past life that look as if they are going to carry on for a good few years yet. Once famous for grasses, Jimi has dropped most of them, as they were taking up too much space for too long without giving enough back. The chionochloa, though, never has an off-day as it is evergreen.

ABOVE
Plants packed into a polytunnel for the winter. Much of the foliage will be removed as the weather gets colder and the risk of fungal disease increases.

OPPOSITE, CLOCKWISE FROM UPPER LEFT
Oreopanax echinops, *Pterocarya stenoptera* 'Fern Leaf, *Senecio cristobalensis*, *Thalictrum* 'Splendide'.

PLANTS FOR

POLLARDING OR COPPICING

2

1

1. *Populus glauca*
2. *Populus deltoides* 'Purple Tower'
3. *Salix magnifica*
4. *Pterocarya stenoptera* 'Fern Leaf'
5. *Populus purdomii*
6. *Ailanthus altissima* 'Purple Dragon'
7. *Eucalyptus* spp.
8. *Hydrangea paniculata*
9. *Paulownia kawakamii*
10. *Paulownia tomentosa*
11. *Populus lasiocarpa*
12. *Populus x wilsocarpa* 'Beloni'
13. *Sambucus nigra* 'Black Lace'
14. Sambucus tigranii
15. Toona sinensis

3

5

4

ASHLEY'S GARDEN CALENDAR

🍃 Foliage ✿ Flowers

	COLOUR	FEB	MAR	APR	MAY	JUN	JUL	AUG	SEP	OCT	NOV
FOLIAGE – HARDY WOODY PLANTS											
Magnolia 'Big Dude'	green	🍃	🍃	🍃	🍃	🍃	🍃	🍃	🍃	🍃	
Populus glauca	green				🍃	🍃	🍃	🍃			
Populus deltoides 'Purple Tower'	purple				🍃	🍃	🍃	🍃			
Populus purdomii	green				🍃	🍃	🍃	🍃			
Pseudopanax crassifolius	purple	🍃	🍃	🍃	🍃	🍃	🍃	🍃	🍃	🍃	🍃
Pterocarya 'Fern Leaf'	green			🍃	🍃	🍃	🍃	🍃	🍃		
DRAMATIC FOLIAGE – NOT HARDY											
Canna 'Musifolia'	green						🍃	🍃	🍃	🍃	
Canna 'Taney'	green					🍃 ✿	🍃 ✿	🍃 ✿	🍃 ✿	🍃	
Colocasia esculenta 'Pink China'	green					🍃	🍃	🍃	🍃		
Colocasia esculenta 'Ruffles'	green				🍃	🍃	🍃	🍃	🍃		
Cyphomandra betacea	green				🍃	🍃	🍃	🍃	🍃		
Ensete ventricosum 'Maurelii'	red				🍃	🍃	🍃	🍃	🍃		
Entelea arborescens	green			🍃	🍃	🍃	🍃	🍃	🍃		
Musa sikkimensis 'Bengal Tiger'	green				🍃	🍃	🍃	🍃	🍃		
Oreopanax echinops	green				🍃	🍃	🍃	🍃	🍃		
Senecio cristobalensis	purple				🍃	🍃	🍃	🍃	🍃		
Telanthophora grandiflora	yellow				🍃	🍃	🍃	🍃	🍃	🍃	
GRASSES											
Chionochloa rubra	light brown	🍃	🍃	🍃	🍃	🍃	🍃	🍃	🍃	🍃	
ANNUALS											
Cosmos sulphureus 'Tango'	orange						✿	✿	✿	✿	
BULBS											
Allium 'Globemaster'	purple				✿	✿					
Allium stipitatum 'Mount Everest'	white				✿	✿					
Allium hollandicum 'Purple Sensation'	purple				✿	✿					
Allium 'Universe'	purple				✿	✿					
Lilium 'Red Flavour'	red						✿	✿			
Narcissus 'Polar Ice'	white			✿							
Narcissus 'Silver Chimes'	white			✿							
Tulipa 'Ballerina'	orange				✿						
Tulipa 'Havran'	purple				✿						
Tulipa 'Jan Reus'	red			✿							
Tulipa 'Merlot'	red			✿							
Tulipa 'Recreado'	purple				✿						

	COLOUR	FEB	MAR	APR	MAY	JUN	JUL	AUG	SEP	OCT	NOV
HARDY PERENNIALS											
Actaea cordifolia 'Blickfang'	white						✿	✿	✿		
Anthriscus sylvestris	white					✿	✿				
Artemisia lactiflora 'Elfenbein'	cream							✿	✿		
Astilbe chinensis var. *taquetii* 'Purpurlanze'	pink						✿	✿	✿		
Astrantia major 'Bo-Ann'	pink			✿	✿	✿	✿	✿	✿	✿	
Eupatorium chinense	pink							✿	✿	✿	
Foeniculum vulgare 'Purpureum'	bronze foliage			🍃✿	🍃✿	🍃✿	🍃✿	🍃✿	🍃✿	🍃✿	
Geum 'Totally Tangerine'	orange			✿	✿	✿	✿	✿	✿	🍃	
Helleborus x hybridus	pink	✿	✿	✿							
Lamium orvala	pink			✿	✿						
Lobelia x speciosa 'Hadspen Purple'	red						✿	✿			
Lychnis 'Hill Grounds'	pink					✿	✿				
Lychnis coronaria 'Gardeners' World'	red					✿	✿				
Lythrum virgatum 'Dropmore Purple'	pink						✿	✿	✿		
Malva 'Gibbortello'	red					✿	✿	✿	✿	✿	
Monarda 'On Parade'	pink						✿	✿	✿		
Persicaria 'Orange Field'	orange						✿	✿	✿	✿	
Primula 'June Blake'	yellow		✿	✿	✿	✿					
Primula elatior	yellow		✿	✿							
Pulmonaria rubra 'Redstart'	pink		✿	✿	✿						
Pulmonaria – blue cultivars	blue		✿	✿							
Roscoea purpurea 'Vannin'	violet							✿	✿	✿	
Sanguisorba tenuifolia 'Pink Elephant'	pink							✿	✿	✿	
Sanguisorba officinalis 'Arnhem'	red					✿	✿	✿	✿		
Sanguisorba tenuifolia 'Stand Up Comedian'	white							✿	✿	✿	
Hylotelephium 'José Aubergine'	red				✿	✿	✿	✿	✿		
Thalictrum 'Elin'	pink				✿	✿					
Thalictrum delavayi var. *decorum*	pink					✿	✿	✿			
BORDERLINE HARDY PERENNIALS											
Dahlia australis	pink						✿	✿	✿	✿	
Dahlia 'Bright Eyes'	pink					✿	✿	✿	✿		
Dahlia coccinea	yellow					✿	✿	✿	✿	✿	
Hedychium greenii	red							✿	✿	✿	
Salvia 'Amistad'	blue						✿	✿	✿	✿	
Salvia buchananii	red					✿	✿	✿	✿		
Salvia confertiflora	red							✿	✿	✿	
Salvia curviflora	blue						✿	✿	✿	✿	
Salvia fulgens	red						✿	✿	✿	✿	
Salvia involucrata 'Bethellii'	pink						✿	✿	✿	✿	
Salvia 'Mulberry Jam'	pink						✿	✿	✿	✿	
Salvia microphylla 'Cerro Potosí'	pink						✿	✿	✿	✿	

FRED'S GARDEN

Ashley's Garden was the first part of Hunting Brook that I felt I really looked at when I first visited, and I'm sure this might be true of many other visitors. However, with the emergence of Fred's Garden that might change. The border that used to run along the side of the broad path up to the house was home to one of Jimi's first plantings and for many years was dominated by big clumps of grasses and late-flowering perennials. The grasses have long since gone and the latest version of the border is clearly all about colour harmonies, but it will be very different to Ashley's Garden. This garden is named after a black and white collie dog that Jimi had for 17 years.

'One of the most exciting parts of gardening here,' says Jimi, 'is completely clearing a border each year and starting with a blank canvas.' Speaking of the planting that used to be present, he says, 'I spent years sourcing the tallest perennials and had a lot of fun growing them, but the new border is going to be very different.' For a start he intends it to be very low as he wanted it to 'flow into the countryside'.

The decision to rework this area came about because Jimi realized that if you are ignoring part of the garden, that's a message to yourself – it needs changing. He started somewhat unconventionally. 'I made a list of things I did not want: no tropical-looking stuff, nothing tall. I wanted to get back the view which had been hidden behind big grasses for years. I didn't want tall plants, bulky plants, big leaves, tropical, blue, yellow, bright red or white flowers. I wanted to give this area its own identity.

'Ashley's Garden has more zing. I don't mind yellow with pink there, I enjoy the clashing colours, but I've had a change of view recently, and my visit to New Zealand might have had something to do with it. There are so few colours there I think it has made me tone things down, which is what I am trying to explore here.' He wanted to look at texture, in particular the contrast of sharp and soft, having spiky plants contrasting with fluffy ones such as *Artemisia stelleriana* 'Boughton Silver' (also known as 'Mori's Form') or the grass *Stipa tenuissima*.

PREVIOUS PAGE
Cactus *Cleistocactus strausii* with *Calendula* 'Indian Prince'. In the foreground is the purple-tinged succulent *Echeveria* 'Duchess of Nuremberg' and beside it the wavy-edged leaves of *Dudleya brittonii*.

ABOVE
What was once the border along the drive being redeveloped as Fred's Border during the winter of 2017-8.

RIGHT
An *Aeonium* cultivar in flower with *Olearia lacunosa* (right) and *Nasella tenuissima*.

He originally thought of agaves for the spiky elements, at least for the summer, but instead has decided to try to get hold of some plants of *Yucca rostrata*, a truly spectacular and reliably hardy species that many of us have admired at Chanticleer. It forms a great sphere of blue-tones leaves atop a steadily-growing trunk but is hard to get hold of, so he may have to be patient.

On the one hand Jimi says he wants 'a quirky mix of eclectic mix of plants' but on the other he intends 'your eye to flow – if anything stops that flow, I need to get rid of it'. I sense a conundrum here that may take a long time working out. Colours lean towards the orange and burgundy, but also with many subtle peachy pinks. There is more silver than has appeared before in the garden, not just the artemisia, but also *Euphorbia rigida* and *E. seguieriana* subsp. *niciciana*. These are all relatively low plants, and the idea is to keep nearly all the planting here lower, except for a minority of taller woody structure plants. There is a particular emphasis on fine foliage, an example being the South African *Inulanthera calva*, a daisy-family plant with bendy stems densely clothed in very fine foliage; Jimi is very fond of it and aims to repeat it through the border. Quite unlike almost anything else we are familiar with (although *Eupatorium capillifolium* comes to mind) it is quintessentially a quirky plant.

Jimi started planning with the repetition plants, which need to have a long season of interest and in some cases, such as the *Stipa* grass, help to create a subtle link to the pastureland on the other side of the road. The long-flowering *Geum* 'Totally Tangerine' is a vividly colourful element which is repeated down through the border, its colour being the element that stands out rather than its form, along with the much more subtle *Linaria* 'Peachy'. In the first year *Calendula officinalis* 'Indian Prince' was used as a repeating annual.

An important influence in the developing colour scheme here has been Fionnuala Fallon, the gardening correspondent for the *Irish Times*, who lives nearby and has a cut-flower farm. 'We even share a polytunnel down the road, so I am watching what she is putting together,' says Jimi. Fionnuala has an eye for colour which tends to explore subtle differences in shade between either intense dark tones or pale ones. Jimi seems to be

PREVIOUS PAGE
The path that runs through Fred's Border is lined with *Artemisia stelleriana* 'Broughton Silver' and the rosettes of *Verbascum bombyciferum*. Dark purple Salvia 'Love and Wishes' and orange *Calendula officinalis* 'Indian Prince' are prominent.

LEFT
Linaria 'Tarte au Citron' with *Geum* 'Totally Tangerine'.

ABOVE FROM LEFT TO RIGHT
Inulanthera calva, *Daucus carota* 'Purple Kisses', *Dahlia coccinea* seedling.

using both in this new border but limiting the actual colours, which will have the effect of making the eye focus on the subtle differences between similarities. In learning about colour and experimenting with it, Jimi says, 'Flower arranging is a good way to work with colour. I do that with classes sometimes, I just go around and grab flowers from the borders and look at shapes too, not just colours. I say to students that I can't teach them about colour as it is such a personal thing – all I can do is show them what I'm into at the time.'

Almost inevitably the most stable element here will be the taller shrubby parts of the mix. Perhaps inspired by the success of the signature *Aralia echinocaulis*, he wants to focus on the vertical: 'Any that I use have to be very see-through and not block the view or stop your eye as you look down the border,' he says. Again, his New Zealand visit has had an impact, with three plants from there being part of the woody structure: *Pseuodopanax* 'Tuatara', *Olearia lacunosa* and *Plagianthus regius*, along with two forms of *Rhamnus frangula* chosen for their finely cut foliage. A key point with these is that he has multiples repeated down through the border, creating a strong sense of year-round rhythm. At a lower level, vertical punctuation is provided by some kniphofias.

At the time of writing, there are some groups of cacti and succulents at the edge of the border, atop a low retaining wall, looking distinctly out of place; the previous year they were in wooden boxes on the decking outside Jimi's house. They surprise, which is probably part of their function, and certainly stimulate conversation among visitors. They prove the point that plants from very disparate habitats can be put together for summer displays; this was, after all, an important part of Victorian horticulture. On a more practical point, they show how these and many other plants can be dug up every year, repotted for the winter, and put back out again next year. Their growth rate will almost certainly be slower than if left to grow undisturbed, but restricting the size of half-hardy plants is rarely a bad thing. The fact that they have to be dug up and replaced every year makes them ideal for experimentation – I suspect that the cacti in particular will always be on the move. This is an experimental border, which will change over the years to come, but one with a real vibrancy and a sense of daring to be different.

ABOVE
Evoking the temporary summer planting-out once extensively used in parks bedding schemes, cacti (*Cleistocactus strausii* predominating) and cultivars of *Echeveria* front Ashley's Border.

RIGHT
Aeonium 'Voodoo' with *Calendula officinalis* 'Indian Prince'.

FAR RIGHT
Salvia 'Love and Wishes' and *Kniphofia* 'Poco Red'.

FOLLOWING PAGE
A young plant of the grass *Panicum virgatum* 'Purple Haze' with *Nasella tenuissima*, next to *Agave attenuata* and an *Opuntia* species (prickly pear cactus). The orange is *Geum* 'Totally Tangerine'.

ANY VERTICALS I USE HAVE TO BE VERY SEE-THROUGH AND NOT BLOCK THE VIEW OR STOP YOUR EYE AS YOU LOOK DOWN THE BORDER

CACTI & SUCCULENTS

FOUND IN FRED'S GARDEN AND A FEW OTHER PLACES!

CONTAINERS OF SUCCULENTS CREATE A TAPESTRY EFFECT. THE CACTUS ARE BROUGHT INTO THE HOUSE FOR THE WINTER AND NOT WATERED FROM SEPTEMBER TILL APRIL

PLANTS FROM VERY DISPARATE HABITATS CAN BE PUT TOGETHER FOR SUMMER DISPLAYS

AEONIUMS ARE WOVEN THROUGH THE PLANTING IN FRED'S GARDEN

FRED'S GARDEN AT SUNRISE

THE HOUSE AND THE WOODLAND GARDEN

To one side of Ashley's Garden, a notice points to the Woodland Garden – but before this path is taken there is a little more around the house to explore. The rear of the house is taken up by what Jimi calls the classroom, for it is where he holds the classes that have always been an important part of the Hunting Brook mission. There are narrow beds outside with some particularly extravagant-looking foliage plants – even for someone with considerable plant knowledge it can be something of a guessing game as to what is hardy and what is not. Around the right-hand side of the house is the site of the new greenhouse.

Jimi trials a different plant group each year, and as with all good trials many are discarded and only the best survive. Trialling plants is part of the key to what makes Jimi a successful gardener and of growing importance in the garden world. He has an insatiable desire for novelty, and something of an aversion to keeping things the same. While plantsman-designer Piet Oudolf has tended to plant up an area and then leave it – parts of his garden are more than 20 years old now with no major intervention other than weeding and cutting back – Jimi likes to constantly edit or replant, endlessly experimenting with new plants and combinations. With some genera, there are so many new varieties coming on to the market that it can be rather bewildering. Setting them aside for a year allows him to make comparisons and get a better idea of a variety's growth than can be gained from siting it in a border with other plants around it. It is not just flowering or foliage quality that is important to him, but also the architecture of plants as they must fit physically into his planting schemes and not overwhelm others around them.

To the side of the house is a space that appears to be a woodland glade, with trees, both deciduous and coniferous, at the rear. We tend to see into their largely bare mid sections, so the shrubs and Japanese maples in front of them are viewed against a dark backdrop. The ground is covered with rather stringy rank grass, above which a few perennials poke their flowerheads. It is clearly one of those experiments in growing perennials in rough grass which has ended as slightly less

than successful, as indeed most of them do. Twice over the next few days of this particular visit, Jimi says something to me about this area 'possibly becoming garden', by which I understand he intends to get rid of it and establish more conventional planting. He says it though as if it were a slightly guilty secret – something he should not be doing.

It's time to return to that sign to the Woodland Garden. We enter by a vast bamboo (a *Chusquea* species) and, in the summer, encounter a green world. In spring this area is very different and certainly feels like the most happening part of the garden. There are bulbs, particularly snowdrops, small daffodils, scillas, erythroniums, and not-bulbs that behave like bulbs: plants such as corydalis and anemones. Hellebores bloom and, somewhat later, trilliums, as well as many more unfamiliar plants. All are quite low level in the way of spring things and look suitably bright and fresh. In the summer there is little in flower here but plenty of fine foliage – endless variations of green, and a wonderful misty drift of the ethereally fine seedheads of *Valeriana pyrenaica*. This is quite a small area and the paths are wood chip (as opposed to gravel out in the sun), so there is a feeling of damp enclosing quietness here. At the far end one can look across into the darkness of the surrounding woodland and see that it quickly drops down into a valley.

This woodsy, shady area is not very different to many other woodland gardens but for one thing – a *Cordyline indivisa* lurking among the trees. This is a New Zealand species that Jimi particularly enthuses about as one of his favourite foliage plants. Unlike the often-seen *C. australis*, this one has magnificent tropical-looking foliage – broad and blue-green with a conspicuous reddish midrib. Jimi reckons it is hardy to –10°C (14°F).

PREVIOUS PAGE
An arrangement with *Dahlia* 'Black Narcissus', pink *Eupatorium maculatum*, white *Actaea cordifolia* 'Blickfang' and grass *Miscanthus sacchariflorus*.

OPPOSITE
Colocasia esculenta 'Pink China' dominates containers of exotic plants on the verandah of Jimi's house.

ABOVE
Yellow-flowered *Chrysosplenium davidianum*, a dense evergreen groundcover, lines the path in the Woodland Garden.

LEFT
In early spring the Woodland Garden is lit up with various forms of *Galanthus nivalis* and *Crocus tommasinianus*.

FOR THE LOVE OF CHAIRS

WHEN THE FURNITURE OF MY HOME WAS SOLD I WAS LUCKY ENOUGH TO GET AN INTERESTING COLLECTION OF CHAIRS

MY HAMMOCK HANGS
BETWEEN MATURE BIRCH
TREES IN THE MEADOW

THESE CHAIRS (LEFT) WERE MADE
BY MY UNCLE OVER 50 YEARS AGO!

A TRIO OF WOODLANDERS

Spring bulbs and other bulb-like plants, known technically as 'geophytes', have historically been dominated by daffodils, tulips and a few other genera. Leafing through the autumn bulb catalogues, one eventually comes to a kind of botanical hodgepodge where all the miscellaneous plants are put. For many of us, this is always the most interesting part of the catalogues. I suspect this is true for Jimi.

Comparing what was included in the 'miscellaneous' section 30–40 years ago and now, it is *Allium* that stands out the most; today's familiar ornamental garlics are a group that was definitely on the sidelines 40 years ago but now seems to have hit the big time. However, it is three genera usually associated with lightly shaded environments that particularly interest Jimi: *Corydalis, Erythronium* and *Trillium*. All three are becoming more widely grown, but none are as straightforward as the bulbs that can sit in paper bags on the shelves of garden centres for months and still perform perfectly next spring. *Corydalis* tubers do not tolerate dry conditions for long but are otherwise very easy plants; *Erythronium* is a genus which does well if conditions are right for it, and at the moment seems to be undergoing a period of fashionability; *Trillium* is notorious for being slow and difficult, but some growers, notably Tony Avent in North Carolina, are promoting some fabulous new selections and proving that the plants can be grown on a large scale commercially. Jimi seems to grow all three very well.

There are more than 400 species of *Corydalis*; most are woodland plants which grow for a comparatively short time in spring – what Americans call 'ephemerals', with soft, delicate foliage and flowers that explore either a range of pinks (although few are really good) or blues, some of which are stunning. Many self-seed easily, and in some gardens this can become almost too much of a good thing; they have, as Jimi notes, 'a tasty seed to encourage "take out" by ants'. If you are sowing them yourself, the seeds need to be very fresh. All die down when stressed by drought, a natural part of their growth cycle, so they tend to grow more vigorously and spread more strongly when their growing season is as long as possible. This is very much true for *Erythronium* and *Trillium* too. The humus-rich soil of Jimi's woodland and the rainfall of Ireland combine here to provide just the conditions they like.

RIGHT
Erythronium 'Pagoda' and purple *Lunaria annua* in the Woodland Garden. The sign reads "I would love to live like a river flows, carried by the surprise of its own unfolding." a quotation from the Irish poet John O'Donahue.

would
like to live

ABOVE
Corydalis 'Korn's Purple', named for Swedish plant hunter and grower Peter Korn (left) and *Corydalis curviflora* (right), reckoned by many to be the best blue-flowered corydalis.

Corydalis solida is the most common species, with pinky-red 'Fire Bird' one of the more intense colours to be had in the genus. *C.* 'Blackberry Wine' is a favourite of Jimi's, a hybrid with fern-like, blue-green leaves and fragrant, wine-purple flowers from May to July. Jimi advises, 'If foliage is cut back after blooming, there is a sparse repeat flowering later in the summer or early autumn. In a cool summer and in soil that does not dry out, plants may bloom throughout the summer.' The ferny foliage of some is worth growing for its own sake, such as *C. temulifolia* 'Chocolate Stars' which Jimi describes as, 'Amazing dark spring foliage but the flowers are a dirty blue.' Of all the blues, some of which can be very good, he particularly rates *C. calycosa*, which is very similar to other blue *Corydalis* but taller and longer-flowering.

Erythroniums are popularly known as dog's tooth violets from the resemblance of their long, thin corms to canine teeth. While not rarities, these are plants which are outside the mainstream and at an interesting and possibly pivotal point in their history as garden plants. For them, Jimi has the kind of visionary plans that few gardeners would dare contemplate.

The lifecycle of erythroniums is like that of most woodland bulbs; in active growth between late winter and mid-spring, with growth, flowering and seed-set all complete by early summer. Jimi says, 'It is important to understand that growth reserves and flowers are laid down the year previous to flowering, so the longer you can keep a plant in leaf the greater the increase in leaf and flowers stems will be the following year; they don't like being dried out. Like *Corydalis*, and even more, *Trillium*, they don't take mechanical harvesting and need careful handling, so it's best to purchase pot-grown plants or bulbs from reputable suppliers who will send them out freshly lifted and packed in something damp.'

I ask Jimi about new varieties of erythroniums. 'Now you are talking,' he says enthusiastically, 'they are all so beautiful, they do really well here – they like the moist soil in spring.' Among the recent introductions, he favours plants such as 'Harvington Snowgoose', with large white flowers, and 'Susannah', yellow. 'I don't need 50 different erythroniums, but I'll probably end up with them,' he says. 'A thousand 'Harvington Snowgoose' is what I want, but it's a slow process. That little Woodland Garden is really a nursery area where I am building up numbers so that I can have lots of erythroniums, and in the future, trilliums, to plant down the Valley. No one has ever seen a hillside of erythroniums in a garden and they work in meadows too, especially the *dens-canis* ones. They divide really easily and in a few years I'll have plenty. In the woodland they get the very best compost and leaf mould, which I dig up in the Valley.'

Of the species and older varieties, Jimi particularly rates *E. tuolumnense* 'Spindlestone', 'a good early yellow and quick to bulk up'; *E.* 'Pagoda', 'huge leaves and yellow flowers, vigorous, so don't let it smother small woodland plants'; *E.* 'Joanna', 'beautiful pink-apricot flowers, this is a must-have'; and *E. revolutum* 'Knightshayes Pink' – 'this is one of the American species with intriguing mottled foliage and gorgeous rosy blooms with prominent yellow anthers, developed at the National Trust garden Knightshayes, in Devon. It increases steadily and naturalizes well.' Knightshayes is one of the few places where I have seen erythroniums successfully naturalized, in moss beneath tree canopies.

Of these three woodland plants, *Corydalis, Erythronium* and *Trillium*, it is the last which has the reputation as the most 'difficult'. Essentially, trilliums are slow to establish, needing time to build up extensive root systems and resenting disturbance. Jimi says, 'They're rhizomatous perennials rather than bulbs, and they don't like drying out. They lose all living roots and have little chance of surviving beyond the first season if they're bare-rooted for any time, so the best way to purchase a plant is to obtain an established potted specimen from a reputable supplier. Ideally, grow them in dappled light – not deep shade – in a soil that is humus-rich and doesn't dry out completely until midsummer. The commoner trilliums on the whole don't worry too much about pH – they dislike heavy clay, but are very hardy and will take intense cold.'

One of Jimi's long-term plans is to propagate his stock of trilliums and plant them up and down the Valley. He advises that they are best divided in growth about three weeks after they finish flowering, so that they will then root out and establish before autumn. Seed has a double dormancy and requires two warm spells and two cold winters to germinate; it will need to be sown directly after collecting. The self-sowing of trilliums in the Woodland Garden here is a wonderful testament to having perfect conditions.

BELOW
Corydalis calycosa flowers later than most, from May to August.

RIGHT
Erythronium revolutum 'Knightshayes Pink', named for the Devon garden where erythroniums have naturalized.

BELOW
The flowers of *Erythronium* 'Joanna' open pale yellow, fade to pink, ending as a mix of yellow and apricot.

ABOVE
Diminutive *Erythronium howellii* from Oregon and northern California.

LEFT
Trillium kurabayashii is thought by many to be among the finest of all trilliums.

MORE GOOD THINGS FOR SPRING

For spring flowers, most gardeners rely on a relatively limited number of genera – in fact daffodils and tulips suffice for many. Some add primulas in the form of polyanthus, while pulmonarias and hellebores have become very popular over recent decades. It is typical of Jimi that he spreads his net much wider; many of the plants he chooses are quieter but have more staying power.

'I'm not an enormous fan of tulips,' he says. 'It irritates me that you can spend so much money on them and then they only flower once, although I do try to plant one area well with tulips every year. I prefer narcissus, but people expect tulips and the classes need to learn about them. June does tulips well – all her beds are raised, which makes a huge difference as they need such good drainage.' Of the few he likes, Jimi cites mostly species tulips such as *T. clusiana* 'Peppermint Stick' and 'Cynthia', *T. whittallii* and *T. bakeri* which will spread by seed and flower every year.

The peak flowering season for alliums is similar to that of tulips, but these are plants with more staying power. Jimi likes *Allium stipitatum* 'Mount Everest', with white drumstick heads, *A. hollandicum* 'Purple Sensation', in his view the most reliable allium of all, and *A.* 'Globemaster', which he describes as a 'spectacular giant mauve globe that follows on from 'Purple Sensation', coming up reliably bigger and better year after year.'

Primulas do not make as big a splash at Hunting Brook as might be expected, and indeed Jimi admits, 'I need to use them better – the dark colours of many of the really nice ones can be lost, especially among bulbs, and indeed bulbs can ruin beds with their foliage after flowering.' We discuss the virtues of the Barnhaven varieties, originally bred in western Canada but now produced by a grower in France, and of the oxlip, *Primula elatior*, which is one of his favourites but does not self-seed sadly. He also mentions 'June Blake', which his sister grew from a tray of cowslips; it flowers for months and is now quite widely available. Another favourite is 'Hall Barn Blue', one of the dwarf, long-lived and strongly spreading cultivars descended from 'Wanda', which was an early variety bred from *Primula juliae*, a species from the Caucasus introduced in the early years of the 20th century.

RIGHT
Dark red *Tulipa* 'Lasting Love' and orange 'Ballerina' in Ashley's Garden in spring with *Pseudopanax crassifolius* silhouetted against the sky.

Of spring-flowering plants, *Epimedium* seems to be of growing importance. This is a genus which has long been popular as a groundcover plant for shade, with flowers as a bit of an added bonus to the tough, heart-shaped leaves. Recent decades have seen many introductions from the Far East, less vigorous but with some extravagantly beautiful and intriguingly shaped flowers. Jimi reckons that some of his Asian forms are bulking up so well that they could perhaps have groundcover potential, including 'Black Sea', which has dark foliage over the winter. He feels that these Asian species may grow better in full sun rather than shade at our latitude. Other epimediums he has that are bulking up well and would make good ground covers include *E.* x *warleyense* 'Orange Queen', *E. membranaceum*, *E.* 'William Stearn', *E. wushanense* (spiny-leaved) and *E. franchetii.*

Among the older groundcover varieties, Jimi rates *Epimedium x versicolor* 'Sulphureum' as a wonderful plant for sun or semi-shade, tolerant of dry soil. He likes using snowdrops, crocus, or species of *Chionodoxa* or *Muscari* planted with them. Of the new Asiatic ones, he recommends *E. membranaceum*, which flowers for about eight months in a nice simple yellow, pink *E.* 'William Stearn', which is also long-flowering, and *E. fargesii* 'Pink Constellation' which he describes as 'one of the most stunning epimediums you could grow, for its nodding pink backswept flowers'. It also has very distinctive long, pointed leaves. Planting epimediums on a slope is a useful tip for helping us to see and appreciate them better as the flowers, with their extraordinary 'jesters' hats are so much better appreciated from below.

Jimi is enthusiastic about many other spring-flowering plants. *Cardamine*, a member of the Brassicaceae cabbage and wildflower family, is one genus he thinks is hugely underrated, containing probably ten species worth growing. He says, '*C. enneaphylla* is a lovely yellow, flowering very early, at snowdrop time; white-flowered *C. heptaphylla* is one that spreads around a bit, a really good garden plant but hard to get hold of unless you go to plant fairs.' Then there is *C. bulbifera*, a pale pink Irish native which 'does spread, a bit of a rogue', while *C. pentaphylla* is also pink-flowered but better behaved. All of these are plants for semi-shade and soils that do not dry out. They are good companions for bulbs and other woodland plants.

Chrysosplenium, a relative of saxifrage, had no garden interest until recently; there is one Irish native, *C. oppositifolium,* which carpets damp, shaded ground in yellow-green flowers in spring. *C. davidianum* is a larger, chunkier Chinese species which Jimi regards as good groundcover that forms a thick carpet over the soil. He says, '*C. macrophyllum* is another Chinese species only discovered in 1996 on Mount Omei. It is a 'must-have' plant when seen in flower, but more like a bergenia in foliage. In a damp and shaded place, such as the bottom of the Valley, it soon sends out long runners to extend its territory and forms sizeable colonies.'

Foliage is often overlooked as an element in the spring garden. In Ireland's relatively mild climate, Jimi says, *'Arum italicum* has leaves which appear in October and look good in the winter garden until March.' The marbled leaves are very distinct, and several cultivars have been selected for good markings: Jimi likes to combine 'Edward Dougal', 'Yarnells' and *A. italicum* subsp. *italicum* 'White Winter' through the Woodland Garden to act as a foil for the snowdrops when they appear.

Most of us only think about grasses later in the year, but Jimi points out that some look pretty good now. *Chionochloa rubra* is one such, a plant that resembles a smaller and much more elegant version of pampas grass. Otherwise it's a matter of the sedges, plants that are easily confused with grasses but are more likely to be evergreen and more tolerant of shade and poor soils. As evergreens they are obvious companions for bulbs and spring flowers.

Jimi recommends *Carex oshimensis:* 'One of the very best evergreen grasses for sun or shade, this makes a wonderful edging plant. *C. morrowii* 'Variegata' is another such. Cut back in spring if they're looking untidy, but only every second or third year as it takes a few months for the plant to recover.'

Jimi's take on shrubs, as we have seen earlier, is quite an idiosyncratic one. Not for him are displays of spring blossom, inevitably followed by a long season of relatively uninteresting foliage – his interest in woody plants is more focused on their foliage value. However, two genera earn his praise for their very early and welcome flowers. He likes *Hamamelis,* especially *H.* x *intermedia* 'Westerstede', which he describes as one of the best winter-flowering shrubs with flowers from December to March, frost- and snow-proof, and *H. x intermedia* 'Pallida', which is his favourite for its bright yellow flowers. He also loves *Daphne bholua* 'Jacqueline Postill', which starts to flower in early January; he says, 'They are very expensive but well worth the spend. Daphnes hate root disturbance, so plant it where you want to leave it!'

OPPOSITE AND ABOVE
Cardamine pentaphylla (opposite) and and *C. heterophylla* (above). Two members of a genus which is much underrated as a source of spring flowers. Some do spread vigorously however.

ABOVE AND RIGHT
Two epimediums bred by Keith Wiley of Wildside in Devon: above, *Epimedium* 'Wildside Ruby' with bright red young foliage; right, *E.* 'Wildside Amber'.

LEFT
E. fargesii 'Pink Constellations' was originally collected by Mikinori Ogisu, a Japanese plant hunter of legendary fame.

BELOW
Epimedium 'Windfire' An American-bred variety with distinct black flower stems.

I ENJOY CUTTING FLOWERS FROM THE GARDEN TO MAKE INFORMAL ARRANGEMENTS FOR THE HOUSE

A BEAUTIFUL ARRANGEMENT BY FIONNUALA FALLON

CUT FLOWERS

IN AND AROUND THE HOUSE

FLOWERS IN JARS FOR TEACHING
MY PLANTSPERSON'S COURSE

FLOATING FLOWER DISPLAYS ARE
A FAVOURITE FOR DINNER PARTIES

SPRING INTO SUMMER

Spring is usually dominated by bulbs, many of which will be gone, or at least past their best, by May. In larger gardens, and in the past generally, this period is associated with flowering shrubs; indeed, the kind of gardens created in Ireland over much of the 19th century became notorious for looking amazing at this time and very dull for the rest of the year, so dominated were they by rhododendrons. Today's greater reliance on perennials, and the simple fact that few of us have much space for many shrubs, means that there may be a slight pause between the bulbs of spring and perennials of summer.

This pause allows time for the plants of woodland-edge habitat to come into the spotlight, and since quite a lot of Hunting Brook is this kind of environment, it seems quite natural to have a lot in flower at this time. The woodland edge is a transitional habitat and many of these plants seem to be happy in either sun or light shade. Such habitats can also be home to more truly woodland plants and some sun-loving species of more open places.

Cardiocrinum giganteum, the giant Himalayan lily, flourishes at Hunting Brook. Flowering spectacularly after some six years from seed, the bulbs then break up to produce offsets which take several years to build up again to flowering size. Jimi's old classmate, Seamus O'Brien, down the road at Kilmacurragh, has large numbers of younger plants of these, organized into generational cohorts so no summer there is without the sight and scent of these extraordinary plants.

May is the time when the last true woodland species flower. These are the plants that grow in locations where they benefit from the light allowed by the absence of foliage on the trees and complete their year's cycle of growth beneath the fresh green shade of young leaves. The moister and more fertile the soil, the longer they will be able to keep on growing, and the less they will be discomfited by the reduced light.

One of the most exciting groups here are the Solomon's seals, or more precisely, a number of genera in the Nolinoideae, a sub-family of the relatively new Asparagaceae. Jimi, along with many other cutting-edge gardeners, sees these as one of the most exciting 'new' groups in temperate-zone horticulture. There have always been a limited number in cultivation, species from Europe or North America. Recent collecting in the Far East, mostly by Sue and Bleddyn Wynn-Jones and

RIGHT
Cardiocrinum giganteum, the giant Himalayan lily flowers majestically and can reach up to 3.5m (11½ft) high.

ABOVE LEFT
Polygonatum falcatum 'Silver Stripe' is a Japanese selection of a slow-growing woodland species.

ABOVE RIGHT AND OPPOSITE
Podophyllum pleianthum has the largest leaves of this woodland genus and, mysterious maroon flowers.

Dan Hinkley, has opened up far more genetic diversity. *Polygonatum verticillatum* is one of Jimi's favourites of the more established species where the leaves are 'arranged like wagon-wheel spokes in tiers up the stems with the muted pinkish bell flowers snugged into the leaf axils, followed by attractive reddish fruit clusters in late summer and early autumn'. These plants, if happy in humus-rich woodland-type soil, will continue to improve year after year until at maturity they begin to look oddly bamboo-like.

Species of *Podophyllum* also make an impact now. Although their flowers and fruit often grab attention, it is their leaves that can stop people in their tracks – dinner-plate size and supported, umbrella-style, by a stalk that connects to the centre of the leaf. 'It is one of my favourite woodland plants, with mottled brown/green leaves and red flowers,' enthuses Jimi. 'Like them all it needs rich moist soil out of direct sun.' Then there is *Maianthemum racemosum* (formerly *Smilacina racemosa*), which is in Jimi's top five plants for shade, good healthy foliage, flowers that smell of almonds and easy cultivation. Finally, *Saruma henryi*, 'Discovered by the Irish plant hunter Augustine Henry,' Jimi announces patriotically, 'a must-have, spring-flowering woodland plant with light yellow flowers and edible leaves.' With its very elegant heart-shaped leaves, this completes a set of plants which really make a striking foliage impact, much more so than any of the species performing so far in the year, and in most cases an impact that should carry on through the summer, at least until August. These plants all benefit from being in soil with a high organic matter to hold moisture; they love a little extra irrigation in dry summers and this can make them among the most spectacular of all shade plants for it seems to encourage large and lush foliage, as it does with trilliums.

Another true woodlander with rather less immediate but still very worthwhile light and feathery foliage impact is deep pink-red *Dicentra formosa* 'Bacchanal'. Jimi finds it an easy woodland plant to grow, although if the soil gets very dry in summer it will go dormant. However, various species and forms of *Dicentra* spread at very different rates, 'Stuart Boothman' is beautiful but very slow,' he says, 'but *D. eximia* is terribly invasive although perfect down in the Valley.' I would only add that dicentra's rate of spread may vary between locations – I never had any luck with any of them!

BRIDGING SPRING AND SUMMER

1. *Anemone coronaria* 'Mister Fokker'
2. *Allium* 'Purple Sensation ' & *Allium* 'Mount Everest'
3. *Astrantia* 'Bo Ann'
4. *Brunnera macrophylla* 'Emerald Mist'
5. *Bupleurum longifolium* 'Bronze Beauty'
6. *Camassia leichtlinii* 'Alba'
7. *Camassia leichtlinii* subsp. *suksdorfii* 'Electra'
8. *Camassia quamash*
9. *Corydalis calycosa*
10. *Dicentra formosa* 'Bacchanal'
11. *Epimedium*
12. *Euphorbia epithymoides* syn. *E. polychroma*
13. *Euphorbia griffithii* 'Dixter'
14. *Euphorbia jacquemontii*
15. *Euphorbia palustris*
16. *Foeniculum vulgare* 'Purpureum'
17. *Geranium phaeum*
18. *Geranium sylvaticum* 'Mayflower'
19. *Geum* 'Farmer John Cross'
20. *Geum* 'Red Tempest'
21. *Geum* 'Totally Tangerine'
22. *Gillenia trifoliata*
23. *Gladiolus communis* subsp. *byzantinus*
24. *Lathyrus aureus*
25. *Libertia grandiflora*
26. *Lunaria annua* 'Chedglow'
27. *Lunaria annua* 'Corfu Blue'
28. *Hesperis matronalis*
29. *Lysimachia atropurpurea* 'Beaujolais'
30. *Maianthemum racemosum* syn. *Smilacina racemosa*
31. *Meconopsis* 'Lingholm'
32. *Meum athamanticum*
33. *Nectaroscordum siculum*
34. *Papaver*
35. *Persicaria polymorpha*
36. *Sinopodophyllum hexandrum* var. *emodi* 'Majus'
37. *Polygonatum verticillatum*
38. *Primula japonica* 'Miller's Crimson'
39. *Pulmonaria*
40. *Sanguisorba menziesii*
41. *Saruma henryi*
42. *Thalictrum aquilegifolium*
43. *Trollius* x *cultorum* 'New Moon'
44. *Trollius* x *cultorum* 'T. Smith'
45. *Valeriana pyrenaica*
46. *Viola* 'Inverurie Beauty'

THE VALLEY AND MEADOW

Turning away from both the extrovert colour of Ashley's Garden and the green intimacy of the Woodland Garden takes us with only a few strides towards a very different world, the Valley, a world also of shade and greens, but on a far bigger scale. As we enter there is a sense of being about to plunge down, as the sides are quite steep; looking across, we can see that the top of the valley opposite is at a similar level to where we are now, so it reads like a great gash in the landscape. While it is overwhelmingly wooded, it is an odd kind of woodland as the canopy of the trees – mostly pine, larch, sycamore, beech and some oak – is very high, so there are few branches to obscure the view down the valley.

Walking down takes us past an area where quite a few of the trees have been cleared, so more light comes to the forest floor. Because of the absence of lower branches it is open and airy enough to be a good place for the long-term plantings which Jimi has been making here. These are very ambitious: large-leaved rhododendrons, viburnums, Japanese maples, species camellias – some with leaves so small it is difficult to believe that they are camellias – and a lot of the ivy family, the Araliaceae: *Fatsia, Kalopanax, Metapanax, Pseudopanax, Schefflera.* This planting of so many Araliaceae is one of Jimi's most important contributions to contemporary horticulture, trialling some impressive foliage plants; for those who do not know them, think of the familiar *Fatsia japonica* and then imagine endless variations of leaves divided into multiples, or with separate leaflets, or webbing between leaflet stems.

The woodland is an important area, because what Jimi is doing here is quite rare; it is a long way off maturity, though. There is a tradition going back to the period of the late 19th and early 20th centuries of landowners planting rhododendrons and other exotica in woodland, but Jimi has used another range of plants entirely. Whereas previous generations indulged a taste for showy flowers, Jimi's focus is clearly on high-quality foliage. Most of these are woodland-edge species which have never been used much in cultivation, and certainly not on such an ambitious scale. As yet there is not much more to be written about this project, but it is somewhere to come back to in future years.

PREVIOUS PAGE
Matteucia struthiopteris is steadily covering the moist valley floor.

ABOVE
The high canopy of the Valley.

OPPOSITE ABOVE
Schefflera delavayi about to be planted out.

FAR RIGHT
Acer pseudosieboldianum ssp. *takesimense* with flowers – a large-growing east Asian species.

WHAT JIMI IS DOING HERE IS QUITE RARE – HIS FOCUS IS CLEARLY ON HIGH-QUALITY FOLIAGE

1

2

3

THE BEST OF

THE ARALIACEAE

AMBITIOUS TRIALLING OF IMPRESSIVE FOLIAGE PLANTS TAKES PLACE IN OPEN, AIRY PARTS OF THE WOODED VALLEY

4

5

1. *Zanthoxylum ailanthoides*
2. *Metapanax delavayi*
3. *Kalopanax septemlobus f. maximowiczii*
4. *Fatsia polycarpa* Needham's form
5. *Schefflera delavayi*
6. *Brassaiopsis mitis*
7. *Aralia echinocaulis*
8. *Neopanax laetus* syn. *Pseudopanax laetus*
9. *Pseudopanax crassifolius*
10. *Schefflera macrophylla*

6

SHRUBS TOLERANT OF LIGHT SHADE

1

IN PLACES, LARGE TREES HAVE BEEN CLEARED TO ALLOW MORE LIGHT TO THE FOREST FLOOR

1. *Cordyline indivisa*
2. *Fatsia polycarpa*
3. *Aucuba chinensis* syn. *A. omeiensis*
4. *Ilex pernyi*
5. *Mahonia eurybracteata* 'Maharajah'
6. *Neopanax laetus* syn. *Pseudopanax laetus*
7. *Osmanthus heterophyllus*
8. *Rubus lineatus*

2

Walking on down, there are areas which are still bare of planting but also a couple of patches where certain plants have done very well indeed and form sprawling, clearly naturalizing, clumps, notably *Dicentra eximia* and the green form of *Impatiens omeiana.* On reaching the bottom of the Valley and crossing over a plank bridge, it is now pretty dark. To one side is a long wooden table, largely covered in moss, with stools made of logs, also mossy. A stoneware flagon sits on the table, its shiny surface the only thing clear of the grasping green fingers of the moss. Any visitor to Ireland who is swayed by the kitschy presentation of the country by parts of its tourism industry inevitably thinks of leprechauns. Jimi confesses that he has not eaten here for a long time and that the seats might be rather damp.

At this point it is apparent that our occasional companions on the journey, an old English sheepdog called Doris and a long-haired dachshund, Billy, are going to be with us for the duration. Having accepted us as members of the pack, they are now going to make sure we are safe for the trip.

From here it is back on up again. This side is steeper and the native vegetation is a little more interesting. There are numerous ferns: male fern (*Dryopteris filix-mas*), broad buckler (*D. dilatata*) and also the Chilean *Blechnum chilense*, well established as a plant that forms extensive patches in damp soil. Further along there are more 'garden' plants establishing themselves in the loose leaf-mould banks where the stream has made a particularly deep cut: *Chrysosplenium macrophyllum* is clearly spreading, a gigantic version of a diminutive mat-forming native plant, *Chrysosplenium oppositifolium* (opposite-leaved golden saxifrage), and *Rodgersia podophylla* has begun to form big patches. There are plenty of other species too, which are more recently planted and not yet fully established: various *Rubus, Rheum* and *Ligularia.*

ABOVE
A place to dine on the valley floor.

OPPOSITE
Dicentra eximia is proving a vigorous spreader in the woodland of the valley.

In contrast to the delicacy of these last two is the muscular chunkiness of certain moisture-loving plants that tend to look at their best now: most are apt to emerge late, often flower in July and in a dry season look a bit scorched and below par later on. They are most definitely not 'support staff', partly because of their size but also because they are the only plants that will flourish in certain moist, shady or inaccessible areas. Rodgersias are ideal for the Valley as they enjoy light shade and the right kind of soil moisture – not stagnant water, but moving through the soil, which makes them good streamside plants rather than pondside. Jimi says, 'They underplant some of the aralias and other things around the house, and down in the woods they are flying along.'

Related to them is *Astilboides tabularis*, with umbrella-like foliage from late spring. *Rodgersia* and *Astilboides* emerge late, usually in May, so there is room for the underplanting of small bulbs or creeping spring-flowering perennials beneath them. They flower in midsummer, with plumes of fragrant, white or pink flowers.

Then there is an odd one out, *Darmera peltata*, which flowers very early, well before the leaves emerge, often as early as March; from April onwards its dinner-plate-sized leaves make a great impact in a moist place. It can spread rapidly, sprouting from strangely shaped woody rhizomes (one garden owner referred to it as the 'dead hamster plant').

The letting loose of garden perennials has been done before by other gardeners, but on surprisingly few occasions. It was at the core of *The Wild Garden*, the best-known work of the great 19th-century garden writer William Robinson (an Irishman, who hastily transferred to the Royal Botanic Garden Society in London after allowing the stoves of his employer's glasshouses at Ballykilcavan to go out on a frosty night, or indeed possibly quenching them, following a quarrel). Robinson wrote this as a polemic rather than on the basis of experience, and the idea of naturalizing perennials has generally been a failure (apart from the unhappy stories of Japanese knotweed and giant hogweed). Grass (in temperate climates at least) generally suffocates perennials, but in shade this does not happen and the possibilities are greatly increased. I am confident that much of what Jimi has planted here will have established itself well within a decade.

Climbing up from here along a sturdy set of steps constructed from larch by Brian Hendrick, a local craftsman who has made all of the wooden furniture here from Hunting Brook trees, we soon reach the top; the trees end here and a dramatic view opens out towards the Wicklow Mountains. But before admiring the view it is worth turning left along the edge of the woodland, for along here is a beech tree with a trunk that curves out over one of the steepest drops in the Valley, forming a wonderfully comfortable place to sit. For Jimi this is a special place where he comes to meditate, to think through difficult decisions, or to see a way through personal turmoil: 'I get a real clear direction in my life when I go to sit under the tree. I always get very clear answers, even though sometimes they are the ones I really do not want to hear,' he says.

OPPOSITE
Rodgersia podophylla is naturalizing in moist parts of the Valley.

ABOVE
An old disco ball shimmers in the forest.

LEFT
Beech trees in the as-yet-undeveloped lower part of the woodland.

FOLLOWING PAGE
Lysichiton americana finds a home beside the stream. This may be invasive in some similar situations.

THE WOODS DO NOT REQUIRE MUCH CARE AND GIVE A GREAT DEAL BACK

While the woods are a scene for some adventurous planting, they are also something much bigger than anything else going on here. Jimi says, 'They make me realize that it is not just me working here, but all the trees and all the creatures too, working together.' The woods do not require much care, but they give a great deal back. 'I don't get to do much in the Valley, just a few days of the year, but it is so important to me; when the gates are shut and the public are gone, it becomes a different place. I feel I can go and connect back to the land.' This connection to the land is fundamental to Jimi: 'No matter where I go to give a talk, I always take a piece of moss, or a twig from one of the trees, in my pocket. I even have it on the podium, as it's an all-important connection to back home.'

The land that used to be the family farm stretches west and south from here. Jimi's border looks out over good pastureland with the bleaker slopes of the Wicklow Mountains further on. There is an ancient standing stone here, and another one on the neighbouring family land. There are also two ring forts further up and a third mock one further down – made, Jimi thinks, as a folly when Tinode House was built in the late 18th century.

After the darkness of the woods and the Valley, the view and the light are welcome. The path sweeps down through a meadow, where Jimi introduced the semi-parasitic yellow rattle (*Rhinanthus minor*) some years ago. The turf is short as a result of the grass being weakened by the yellow rattle, plantain flourishing instead. There is a smattering of perennials here, and Jimi is thinking of introducing more to create some form of perennial meadow. Walking down to the bottom of the hill and the stone-faced embankment of the mock ring fort, we encounter a couple of large garden perennials that appear to guard re-entry to the woodland from the more open landscape: yellow flowered daily family *Telekia speciosa* and the knotweed *Aconogonon* 'Johanniswolke', a reminder that this is somewhere where anything is horticulturally possible. Following the path soon completes our journey by bringing us back to the point where we entered the Valley.

LEFT
Beech trees in the woodland are rewarding in all seasons.

ABOVE
Valeriana pyrenaica seeding itself into grass adjacent to the Woodland Garden where it is slowly spreading.

PART 3

PLANTS AND PLANTING

PRACTICAL WORK

Maintenance in Ashley's and Fred's Gardens is obviously quite intense, but it happens in two concentrated bursts. Planting the tender plants in May is a major task, as is taking them back out again at the end of October. During the summer there is constant deadheading, but not much else to be done. Above all, there is no staking; Jimi simply does not bother with plants that need it. Indeed much here has the relatively airy quality of plants that do not build up massive heads of top-heavy foliage or flower. Making decisions about how to change and improve the planting is a pretty constant task, however. Jimi's way of recording decisions is to make a note on a label (usually 'dig out', he says) and tie it to the base of the plant, which he finds much better than taking photographs and making notes.

Planting out is a complex business, as design succession and size have to be worked out, finding spaces among the permanent perennials and popping in the dahlias, salvias and tender foliage plants that have been overwintered under cover. Jimi's fundamental design principle is that 'repetition is very important, particularly of a strong colour, like this year's bright orange *Cosmos sulphureus* 'Tango' through Ashley's Garden'. This cosmos is an annual, grown in plugs in a polytunnel from a March sowing and then dotted through the border to create an underlying colour rhythm, 'I start them in plug trays in a propagator in February and then pot them into 9cm (3½in) pots so that when we plant them out they take off quickly. They are spaced at every 50cm (20in). The numbers of these rhythm annuals can be large – 700 *Calendula officinalis* 'Indian Prince' were grown for the first year of Fred's Garden.

The October dig-up is an even more involved operation than the May planting as it means having to make decisions about the best way to overwinter plants in some very confined spaces. 'Until recently I managed with just a small polytunnel in some shade but I've now upgraded and have 3 polytunnels and a new greenhouse which is my pride and joy,' Jimi explains.

Some dahlias are left in the ground over the winter, mostly the singles, although this often results in them flowering late. Others are dug up and stored inside. Jimi shakes off the soil, lets them dry off on the floor of the classroom,

ABOVE
Ensete plants being lifted for the winter (left) and dahlia tubers in crates ready to go into the classroom for the winter. They won't be watered until March.

RIGHT TOP
Collecting berries of *Viburnum betulifolium*, inset: dahlia seeds.

RIGHT BOTTOM
Doris with plants protected for the winter (left) and salvia cuttings taken in perlite before the first frost.

FOR SUCCULENTS, WET CONDITIONS ARE THE KILLER – THEY ARE NEVER WATERED IN WINTER

then they are put in crates until February; there is no heating in there, and they are not wrapped or put in compost. At the end of February they are potted up to start growth in the tunnels, so that they are in strong growth by May and flowering by the end of June. 'You can sow the seed in the warm in February too,' Jimi says. 'It comes up like mustard and cress.'

South American salvias vary greatly in hardiness. 'I leave out a lot of the small macrophylla types,' Jimi says, 'but I rely on the big ones too much to risk losing them, so they are kept in the tunnels.' However, their winter management is not easy; these are plants that naturally have only a minimal dormant period, but winter growth is not wanted as it will tend to be very weak. Rotting is a big problem and dampness kills them really easily. Jimi tried to keep the tunnels warm but decided that was not helping, so instead he aims to ventilate, keep the air moving and ensure the plants are clean of dead foliage. While he formerly used the finer brands of horticultural fleece he never liked them, as they cling to plants and can make them wet. Now he uses a product called Enviromesh® as a blanket to help keep frost off. This is a relatively stiff material which holds itself well back from foliage, thus helping to keep pockets of warmer air around plants on frosty nights. He also uses it on select young plants outside, tying it around canes to make insulating wigwams. It is not often I feel in a position to offer Jimi advice but on this occasion I could. 'Get a cheap dehumidifier for every tunnel – it does wonders in avoiding high humidity and fungal diseases and keeps air moving.'

Tetrapanax papyrifer is grown in the Woodland Garden, Valley and in containers and is widely admired. One of the less hardy of the Araliaceae family of which Jimi is so fond, its expansive golden-haired foliage has made it very popular. In milder regions it can be almost evergreen, but at Hunting Brook the plants were treated as a deciduous shrub. Jimi advises that if you live in a particularly cold area it is a good idea to protect the trunk in winter with straw or fleece.

Aeoniums and other succulents are potted up and kept in a tunnel. For them it is wet conditions here that is the killer, and they are only watered once or twice during the winter. Cannas, bananas and other foliage plants are put into pots, but the bananas are a real problem, especially *Ensete* bananas (important visually for their big leaves). 'They rot very easily,' says Jimi. 'I have had to try to keep them at an angle to drain the water out of the core of the plant. This year I am going to pot them up, keep the leaves on them and do nothing.' The largest bananas are *Musa basjoo*, a relatively hardy species, so this is left outside and wrapped in Enviromesh® secured with chicken wire. One clump of this species came from a nearby garden; Jimi says, 'I got it from Tallaght, in the modern part of Dublin you drive through to get here. A guy rang me up one day to ask if I wanted it, otherwise he would take it to the dump. I rushed down there and got it into the car, hanging out the back, to bring it home, so I call it the Tallaght banana.'

These are the only plants that Jimi feeds through the summer; 'I feed them everything I have got, manure or seaweed feed – I do the same with the cannas. Feeding them helps to make up for the lack of summer heat, I think.'

On the subject of cannas, Jimi says that he will also lift plants as undivided solid clumps and overwinter them in a dry, frost-free place. In spring they may be divided. He particularly flags up the dangers of plants becoming infected with a virus, initially recognized by pale-coloured spots and streaks on the leaves, distorted or crinkly leaves and later, by stunting and dead rust-coloured streaks in the foliage throughout the plant. There is no cure for virus diseases, and he advises that plants that are obviously diseased should be dug up and destroyed. Buying from reputable nurseries rather than from garden centres is an important first step to avoiding disease, particularly important if you garden organically as Jimi does.

OPPOSITE
Plants are squeezed into a polytunnel for the winter, bananas (*Musa* and *Ensete* species) are laid on their sides to allow water to drain out of the crowns of the plants; they are then potted up and kept dry until March.

JIMI'S PLANT DIRECTORY

TREES AND SHRUBS

ACER JAPONICUM VAR. *ACONITIFOLIUM* ▼

One of the best small trees for autumn colour on the edge of the garden meadow at Hunting Brook. As with all acers, it needs protection from the wind.

ACER JAPONICUM 'AKI-HI'

A small deciduous tree similar to *A. japonica* var. *aconitifolium* but with much larger and more distinctive leaves – round and fresh green, blushed with purple-brown. The autumn colour ranges from orange to red to purple. It is a real show-stopper as you enter the Valley at Hunting Brook.

ACER *PALMATUM* 'OSAKAZUKI' ▼

When I was starting the garden at Hunting Brook, I rescued this tree from Mount Usher Garden in Wicklow, where a new car park was being made. Situating it in the middle of the garden meadow was a happy accident. With its spectacular deep crimson autumn colour, it is the most photographed tree at Hunting Brook.

ACER PSEUDOSIEBOLDIANUM SUBSP. TAKESIMENSE

An exceptionally beautiful small tree, this has finely dissected, lobed leaves. The autumn colour is yellow and infused with orange and metallic red.

AESCULUS HIPPOCASTANUM F. *LACINIATA*

Slow-growing with wonderful dark green, deeply-lobed deciduous leaves which turn a buttery yellow in the autumn. I keep it pruned to 2.1m (7ft) to fit into the borders. This quirky foliage plant always surprises and delights visitors to the garden.

AILANTHUS ALTISSIMA 'PURPLE DRAGON'

I recently saw this growing in Dan Pearson's home garden, where it looked spectacular with wine-coloured pinnate leaves. Mine is still a relatively young tree, but once it matures I intend to experiment with pollarding it.

ANISODONTEA CAPENSIS 'EL RAYO' ▼

A member of the mallow family, with aromatic leaves and bright pink flowers with a purple centre, it blooms all year round at Hunting Brook. One flowered non-stop for three years until the snows of winter 2018 flattened it.

ARALIA CALIFORNICA

This huge architectural plant growing up to 2.4m (8ft) is planted near my nursery. It is endemic to California and Oregon and is found in moist, shady woodland and along the banks of creeks and streams. It has lush-green pinnate foliage and racemes of attractive white flowers followed by black berries. I also grow it in shade in the Valley where it does not flower but has impressive foliage.

ARALIA ECHINOCAULIS ▼

I collected the seed of this in Hubai, China, while on a plant-hunting trip with Seamus O'Brien from Kilmacurragh Botanical Gardens in 2003. Planted close to the house, the trees create a light canopy over the terraced perennial beds without blocking too much light – a most unusual and almost tropical effect. This has become one of Hunting Brook's signature plants.

ARALIA VIETNAMENSIS

Some plants are worth the trouble of bringing them in and out of the garden each year, and this is one of them. It is a tender aralia with glorious golden velvet leaves and a spiny trunk. I incorporate it into different planting schemes to suit my needs.

BETULA ALBOSINENSIS 'BOWLING GREEN'
Near the car park, I have planted a grove of various birches grown for their unusual bark. This has rich chestnut outer bark peeling to reveal a paler silvery under layer.

BETULA ALBOSINENSIS 'CHINA ROSE'
This Chinese red birch is a real beauty with more solid red-coloured bark than others. The bark is smooth and peels in sheets.

BETULA DAHURICA 'MAURICE FOSTER'
A rare birch with eye-catching peeling, papery bark. The outer layers are pale creamy pink, with darker reds and browns underneath.

BETULA ERMANII 'HAKKODA ORANGE'
This unusual and graceful tree has orange peeling bark. The heart-shaped leaves first appear orange, later fading to lime green. It is a very special tree, well worth sourcing.

BETULA UTILIS 'BUDDHA'
A Himalayan birch with a subtle coppery-brown outer layer of bark and shiny white underneath.

BETULA UTILIS 'DARK-NESS'
This birch has dark brown peeling bark with conspicuous white lenticels, looking just as good as any *Prunus serrula*.

BETULA UTILIS 'NEPALESE ORANGE'
One of the most beautiful birches, with rich orange bark. Its slender growth casts only a light shade. This is one of the new birches from Pan-Global Plants in the UK, introduced for their interesting barks.

BOCCONIA FRUTESCENS ▼
I saw this *Bocconia* happily growing all over hillsides in Costa Rica. It will not grow as large in the Irish climate and I have to bring it inside every winter. However, after a hard pruning in spring, its magnificent foliage emerges again for the new season. The leaves bear a remarkable similarity in appearance to *Macleaya*. If I had to pick a favourite foliage plant, this would be it.

BRASSAIOPSIS FATSIOIDES ▼
When I talk about foliage that excites me most – well this is it. *Brassaiopsis* have the most fascinating foliage of any group of plants that I grow in Hunting Brook. This particular one has large palmately lobed leaves which give me the drama I like for container planting. I can never grow this outside in winter but how could I not have a leaf like this!

BRASSAIOPSIS MITIS ▼
This has to be the funkiest leaf of any foliage plant. I was after it for many years and managed to get it last year from Christian Monnet in France. It is not one that can be left out for winter.

CAMELLIA X WILLIAMSI 'NIGHT RIDER'
I am introducing species camellias into the Valley. This one has young leaves that are shiny purple-red, followed by semi-double deep wine-coloured flowers with a velvety texture.

CORDYLINE INDIVISA ▼
Commonly known as the mountain cabbage, this plant is normally temperamental to grow. However, they do very well at Hunting Brook and give a tropical feel to the Valley with their blue sword-shaped foliage and defined midrib, which is often tinged with red and orange. They are hardy at Hunting Brook but do need a sheltered spot in part-shade.

TREES AND SHRUBS

CRYPTOMERIA JAPONICA ARAUCARIOIDES GROUP
This quirky, slow-growing conifer has long, green rope-like foliage that makes it a great specimen plant. It looks a bit like *Araucaria araucana* (monkey puzzle tree). At Hunting Brook it is combined with *Alstroemeria* 'Orange Glory' to great effect.

CRYPTOMERIA JAPONICA 'RASEN-SUGI'
With their columnar form, these unusual conifers are used as gateway plants into the Valley at Hunting Brook. Each branch has short, rubbery needles that twist themselves around the branch. In fact all parts of this plant twist, creating an unusual effect.

DAPHNIPHYLLUM HIMALAYENSE SUBSP. *MACROPODUM*
This evergreen shrub with large rhododendron-like leaves is repeated throughout the border and down into the Valley garden. It is valued for its structure and foliage – the flowers are small and insignificant. If there is a late frost or a cold spell, the new foliage can be damaged but quickly recovers.

DECAISNEA FARGESII
I found the seeds of this tree in China and grew them at Hunting Brook. If pruned correctly into a multi-stemmed shrub it will give you light, airy foliage which is perfect for a small garden.

EUPHORBIA X *PASTEURII*
A relatively new cross between *E. mellifera and E. stygiana*, this is a superb evergreen at Hunting Brook. On warm, sunny days, the seeds explode and often hit garden visitors in the face. It can be pruned hard back in spring to produce attractive fresh growth.

EUPHORBIA STYGIANA ▼
The best evergreen shrub in Hunting Brook, *E. stygiana* produces a perfect dome shape. The snow flattened it one year, which is was a good thing as it was getting huge and needed pruning – it regrows from the base. In my opinion it is far superior to *Euphorbia mellifera*.

FATSIA POLYCARPA ▼
This new *Fatsia* is from Crûg Farm Nursery in Wales. The seeds were collected in Taiwan, China where it grows at high altitude, forming a branched small tree or shrub and reaching 3.5m (11½ft) tall. This is one of the best introductions to my garden, creating an interesting small tree in the Woodland Garden. I raised its canopy as it was growing.

FATSIA POLYCARPA LARGE LEAF FORM
This is *Fatsia polycarpa* on steroids! I first saw this growing in Tregrehan, Tom Hudson's garden in Cornwall, and it made a huge impression. I am nursing a young plant down in the Valley garden and hope it will become as spectacular as Tom's.

FATSIA POLYCARPA NEEDHAM'S FORM
This recently introduced form of *Fatsia polycarpa* is acclaimed for its deeply lobed leaves. I planted a group of them in the Valley only to soon find that a neighbour's sheep had eaten them to the ground! After a few years, they are finally bouncing back and forming small shrubs.

FRANGULA ALNUS 'FINE LINE' SYN. *RHAMNUS FRANGULA* 'FINE LINE'
A funky-looking shrub, it is perfect for a small garden with its lacy, fern-like foliage and columnar habit. I will be adding more of these into Fred's Garden.

FRANGULA ALNUS 'ASPLENIFOLIA' SYN. *RHAMNUS FRANGULA* 'ASPLENIFOLIA'
This is a columnar-shaped shrub with fine fern-like leaves similar to *Acer palmatum* var. *dissectum*. I am using this through the long border to add vertical interest, repeated through the bed.

HYDRANGEA ASPERA SUBSP. *STRIGOSA* FROM GONG SHAN, CHINA
I obtained this superior form of *Hydrangea aspera* from Billy Alexander in Kells Bay in south-west Ireland. It has large green felted leaves with maroon backs and pale lilac lacecap flowers.

HYDRANGEA PANICULATA 'VANILLE FRAISE'

One of my favourite paniculata hydrangeas, with loose, fluffy, pyramid-shaped flowerheads forming on the tips of graceful arching branches during midsummer. Pruned to a framework every spring, they are guaranteed to create beacons of light in any woodland garden.

HYDRANGEA SCANDENS SUBSP. *CHINENSIS* F. *ANGUSTIPETALA* 'GOLDEN CRANE' ▼

A Dan Hinkley selection from Sichuan Province in China, this plant is also known as *Hydrangea angustipetala* 'Moon Long Shou'. I am building up a collection of hydrangeas in the Valley, mostly lacecaps and serratas. This one's lacecap flowers start white and age to a soft yellow from late spring onwards, making it one of the earliest hydrangeas to flower.

KALOPANAX SEPTEMLOBUS F. *MAXIMOWICZII* ▼

A Japanese tree with huge, drooping, deeply lobed leaves on wonderful spiny stems and trunk. This fits with the exotic look I am creating, and it is totally hardy.

LINDERA TRILOBA ▼

I have always considered this the perfect shrub for a small garden. It is a multi-stemmed deciduous plant, originally from Japan. Its habit is open, with a unique leaf structure, having three pointed lobes reminiscent of a hoofprint. It is planted in the Valley, where it is very slow-growing.

LINNAEA YUNNANENSIS SYN. *DIPELTA YUNNANENSIS*

A deciduous shrub with an arching habit and cream-coloured orchid-like flowers with orange markings in early summer. I prune it to show off its peeling white bark.

LOMATIA FERRUGINEA

This valued plant is an excellent evergreen foliage tree. It has fern-like leaves and reddish-brown velvety stems with orange and scarlet flowers. Lomatias need lime-free and well-drained soil and shelter from cold winds. I obtained this plant from Mount Usher Gardens when I was starting Hunting Brook.

MAGNOLIA SUBSP. *ASHEI X MACROPHYLLA* SUBSP. *DEALBATA*

While on my travels visiting some of the best gardens in the USA, I was hugely impressed with the large-leaved magnolias and was inspired to source some and grow them in the Valley. This particular cross creates real drama with its 90cm (3ft) long leaves.

MALUS SIEBOLDII

This rare semi-weeping crab apple from China, Japan and Korea has masses of fragrant pink flowers in spring. Later it produces clusters of small yellow or pale red fruit with good autumn foliage. The leaves have a red midrib as they open in spring. This is a tree for all seasons.

METAPANAX DAVIDII

An impressive member of the Araliceae family and one of the hardiest at Hunting Brook, surviving at -15°C (5°F) with no problems. It is very similar to *Metapanax delavayi* (see below), although the leaves of *M. davidii* are more palmate. I feel it deserves to be grown more.

METAPANAX DELAVAYI

Growing on the edge of the Woodland Garden, this small tree was a generous gift from Helen Dillon. Many of the Araliaceae family are borderline tender, but it has survived Hunting Brook's harshest weather. With long, arrow-shaped, weeping leaflets, this is an exceptionally graceful plant.

MONTANOA BIPINNATIFIDA

This chrysanthemum tree was a gift from Carmel Duignan, one of the greats of Irish gardening. It has dramatic leaves which are deeply indented and feel like a cat's tongue, but has never produced flowers at Hunting Brook. It is brought into the tunnel for the winter.

NIPPONANTHEMUM NIPPONICUM

Montauk or Nippon Daisy is a member of the Asteraceae family and was formerly named *Chrysanthemum nipponicum*. It flowers in October/November and is the last and best autumn-flowering plant at Hunting Brook. It bears clean white daisies on semi-woody stems, a welcome sight at a time when many other garden plants have grown tired.

TREES AND SHRUBS

OLEARIA LACUNOSA ▼

This silver-leaved evergreen shrub from New Zealand is repeated to great effect throughout Fred's Garden created in 2018. I love it for its quirky lance-like leaves which bring another dimension to the border similar to the *Pseudopanax* cultivars grown nearby. Growing shrubs like this one with sparse foliage allows me to get another layer of interest through my perennial planting without blocking light.

OPLOPANAX HORRIDUS ▼

I first saw this in Dawyck Botanical Gardens in Scotland, where it had obviously grown for years outside in all weather. I anticipated correctly that it was hardy and could be used to create a tropical effect in the Valley and was proved right. It has massive palmate spiny leaves and stems and clusters of bright red berries in summer.

OREOPANAX CAPITATUS

I saw this tree in a cloud forest in Costa Rica; I was driving up a steep road and suddenly started recognizing all my favourite plants, including *Schefflera* and *Oreopanax*. It has large shiny leathery leaves similar to a giant ivy tree.

OREOPANAX ECHINOPS ▼

This *Oreopanax* has golden new palmate leaves which are to die for. It is not hardy at Hunting Brook and so comes into the house for winter.

OREOPANAX INCISUS ▼

When I first came across these in a garden centre I suspected they would not be hardy, but I was drawn to them for their spectacular deeply lobed palmate leaves with new purple growth; they can reach 60cm (2ft) in length. Time will tell how hardy they are at Hunting Brook.

OREOPANAX XALAPENSIS

Oreopanax are tender and would never survive winter outside at Hunting Brook, but the large palmate leaves make it worth the trouble of bringing them in for winter and out for summer. *O. xalapensis* does not have purple new growth but the mature leaves add great drama to the exotic garden.

OSMANTHUS HETEROPHYLLUS

I keep this as a short, dense shrub which is grown in semi-shade. So far, it appears to be doing very well. It has deep green, shiny, holly-like leaves and is ideal for modern gardens.

PAULOWNIA TOMENTOSA

Paulownia are pollarded throughout the garden to create enormous furry leaves up to 90cm (3ft) wide in the borders. This deciduous tree is hardy at Hunting Brook and is really useful for creating tropical drama.

PENTAPANAX LESCHENAULTII ▼

Another member of the Araliaceae family, this is totally hardy. At Hunting Brook it grows as a free-standing upright shrub, but it also grows epiphytically in the wild. It has spectacular golden colour in autumn and large round seedheads.

PINUS BHUTANICA ▼
This is one of the most beautiful of all conifers and I was delighted when I finally found it and planted it in a sheltered area in the Valley. It is similar to *Pinus wallichiana* in that it has long, strongly drooping needles. Sadly, it is on the red list of endangered species.

PINUS MONTEZUMAE 'SHEFFIELD PARK'
I first saw this at RHS Wisley outside London and knew I had to have it for its impressive long blue needles. It now holds pride of place in the Valley.

PINUS PARVIFLORA 'RYU-JU'
I planted three of these miniature pines in a well-drained, sunny spot at the entrance to Hunting Brook. Growing out of a carpet of *Celmisia mackaui*, they create a bonsai effect on a dry slope. It is time to start growing more funky conifers like Matthew Pottage does at Wisley and Fergus Garrett at Great Dixter.

PLAGIANTHUS REGIUS
I cannot understand why this graceful, slender tree is not more popular in our gardens. It is similar to birch but more columnar and compact in form. I use it in both the Valley and in Fred's Garden to create vertical height.

POPULUS DELTOIDES 'PURPLE TOWER' ▼
This is another specimen tree that is used for pollarding in the borders. Doing this each year creates large, glossy, deep wine-coloured leaves and keeps the tree to scale in the garden.

POPULUS GLAUCA ▼
A rare tree with large, rounded, young wine-red leaves with pink undersides, it always creates a reaction with visitors when I tell them it is a *Populus*. I pollard the tree in spring to achieve the best effect.

POPULUS LASIOCARPA
Another spectacular foliage plant that can be pollarded to provide a tropical effect from a hardy plant. I use this in permanent shrub planting, combined with bamboo, to great effect. These are used in Dublin Zoo by the very talented Stephen Butler.

PSEUDOPANAX CRASSIFOLIUS ▼
I have a good few of these at various ages. I was very lucky to be given two mature plants by my friend Bernard Hickie, who is a master of designing with foliage plants. The dark, almost black leaves are stiff and leathery with a prominent central rib, all growing downwards from a central stem. This pseudopanax is completely hardy and is becoming a really important addition to the strange foliage plants I use through the perennial beds for structure. They do not block the light.

PSEUDOPANAX LAETUS
An attractive evergreen shrub with distinctive large, shiny leaves, used as the repeat shrub to create a lush understorey in the Valley. It can also be used to great effect when it is carefully pruned as a specimen shrub in a large container.

PSEUDOPANAX LESSONII 'MOA'S TOES'
A recent cultivar introduction that is now readily available, it has leathery deep green leaves with a red tint to them. I have planted 15 of these in a group in one of the borders. I love their quirky, upright habit – they add a sense of fun to the garden.

TREES AND SHRUBS

PSEUDOPANAX LESSONII 'TUATARA'
Another plant for vertical see-through interest in the new border, it mixes well with an underplanting of *Stipa tenuissima*. This evergreen shrub with fine-cut leathery leaves is easy to grow in full sun.

PTEROCARYA STENOPTERA 'FERN LEAF'
Chinese wingnut is my favourite tree to pollard in spring. Usually I pollard trees to get large round leaves but this tree gives the most unusual fern-like foliage which adds another dimension to the perennial and tropical planting. It provides a lovely yellow to the garden's palette during autumn.

RUBUS CALOPHYLLUS ▼
I always plant rubus with caution in the Valley and keep a close eye that they do not spread. This newly discovered one, which makes a medium-sized arching shrub, is worth growing for its pleated leaves and silver undersides. I have planted it on a bank so that the undersides can be seen.

RUBUS LINEATUS
This is a striking plant with pleated, semi-evergreen leaves. They are silky smooth green on top with silvery undersides and are produced on arching stems. A fantastically elegant foliage plant, it is as good as any schefflera and hardy. It tends to sucker around, but this is easy to control.

RUBUS SQUARROSUS ▼
I have such a love of vicious-looking plants that I am considering making a vicious border at Hunting Brook. A leafless vine with conspicuous yellow thorns – doesn't it sound handsome!

SALIX HASTATA 'WEHRHAHNII' ▼
This slow-growing willow with masses of silvery catkins in spring and attractive golden anthers is perfect for a small garden when pruned into a good shape.

SALIX MAGNIFICA ▼
When discovered, *S. magnifica* was thought to be a magnolia from its huge leaves, which are leathery and glaucous. I pollard it at Hunting Brook every spring.

SALIX MOUPINENSIS
Slow-growing for a willow and easy to prune into a shape suitable for a small garden, it is similar to *Salix fargesii*. I prune this each spring to keep an open goblet shape.

SCHEFFLERA DELAVAYI ▼
This is my pride and joy! One of the finest broad-leaf evergreen introductions in recent years, it has wide umbrellas of lobed, oak-shaped leaves with incredible speckled gold new growth. It is the most spectacular foliage plant in Hunting Brook. I first saw this grown at Chanticleer Gardens in the USA, where it survived their cold winters. It is thriving in the Valley at Hunting Brook, where it gets excellent drainage, shelter and a south-facing slope.

SCHEFFLERA KORNASII ▼
A great find from Fansipan mountain in Northern Vietnam, with distinct whorls of leaf with dark purple petioles. This new schefflera has joined the schefflera glade in the Valley. The new growth is silver and creates quite a show.

SCHEFFLERA MACROPHYLLA
When this schefflera was first released for sale at Crûg Farm Nursery there was almost a stampede of Araliaceae nerds, each craving this new introduction, no matter the cost. For the record, yes, I was one of them! I grew it for a few years, bringing it in and out until one summer it decided to die for no apparent reason. I have started again with small seedlings which I hope will survive in the Valley. This schefflera is one of the most spectacular foliage plants with paddle-shaped leaves growing up to 90cm (3ft) long; the new growth is covered in golden indumentum.

SCHEFFLERA RHODODENDRIFOLIA ▼
I have a group of these planted in the Valley and they are thriving. It is a good choice of schefflera for a cooler climate.

SCHEFFLERA SHWELIENSIS ▼
Another scheffie for my glade in the Valley. I happily received the plant from Nick Macer and love its glossy leaves. Being new to me and Hunting Brook its hardiness is unknown, so I brought mine in for the first winter but plan to leave it out the following winter.

SCHEFFLERA TAIWANIANA
A wonderful schefflera that is hardy at Hunting Brook. I was delighted to see this for sale in my local nursery, Mount Venus Nursery. Scheffleras are easy to propagate from cuttings or ripe seeds, which I have done. Unfortunately, my schefflera seeds tend not to ripen due to cool Irish Septembers.

SENECIO CANNABIFOLIUS
As the name suggests, this is cannabis-like in appearance – a tropical-looking plant that is totally hardy. It has yellow ragwort-type flowers in late summer. It seeds around a little but this never creates a problem.

SENECIO CRISTOBALENSIS ▼
I originally got this from Helen Dillon and it is a real favourite here at Hunting Brook. It is a tender foliage plant with dinner plate-sized velvet leaves that are lobed in shape and dark green on top with rich purple underneath. I plant this out into the borders for the summer and bring it back into the unheated tunnel for the winter. I cannot imagine my garden without this plant.

SONCHUS ARBOREUS
A tree dandelion that is fairly hardy at Hunting Brook. It has a woody stem up to 1.5m (5ft) with rosettes of serrated leaves at the end of each branch and sprays of yellow dandelion-like flowers. I saw these growing wild in Tenerife in the Canary Islands while I was hiking there a few years ago.

SONCHUS PALMENSIS
I have a soft spot for dandelions, including the weed which I add to my green juice in the spring. This dandelion tree comes from the island of La Palma in the Canaries and forms a trunk of up to 2m (6½ft) with silvery, green fern-like leaves and yellow dandelion flowers. I bring this into the classroom each winter.

TREES AND SHRUBS

SORBUS HARROWIANA ▼

I grow this sorbus for its thick, leathery foliage unlike any other in the genus. I have planted it in the Valley with *Pseudopanax* and ferns.

SORBUS HEDLUNDI ▼

Probably the most handsome of the whitebeams at Hunting Brook, it has large silvery-white leaves and a rusty dusting underneath which is most attractive. It is very slow-growing, and I imagine it would be suitable for a small garden with some pruning.

SPHAERALCEA 'CHILDERLEY'

I have developed a new interest in the Malvaceae family. There are many new ones on the market and they are incredibly long-flowering. 'Childerley' has soft apricot-orange malva flowers from June until September with grey-green leaves and upright bushy growth. It needs full sun and is hardy in well-drained soil; it flourished in our dry summer of 2018.

SPHAERALCEA 'NEWLEAZE CORAL' ▼

An excellent drought-tolerant plant that did well during our long hot summer of 2018, flowering non-stop from early summer through to late autumn with beautiful coral-coloured flowers.

STYRAX FORMOSANUS VAR. *FORMOSANUS* ▼

This is the most beautiful of all the styrax with its pendulous star-shaped flowers and the most captivating smell of jasmine.

TETRACENTRON SINENSE VAR. *HIMALENSE* ▼

Foliage is always a key consideration for shrubs in the garden, more so than flowers, which are often fleeting. This is a particularly good form of *T. sinense*, with its long, glossy, slender-tipped heart-shaped leaves and distinct red petioles.

TOXICODENDRON VERNICIFLUUM SYN. *RHUS VERNICIFLUA*

Another of the China collection where they were grown as a crop. The sap can be collected from the plants and processed into a lacquer. The plant provides the same dappled shade effect as the aralias; however, it is worth noting that the sap of this plant can severely burn skin. Handle with care!

TRACHYCARPUS PRINCEPS

At 305m (1000ft) above sea level, Hunting Brook is probably not the best place to start collecting palms. However, I could be tempted to experiment. This is one of the most beautiful of the genus, with slight blue colouration of the leaves. I have it planted in the Valley and it is sitting there trying to decide if it wants to grow!

TROCHODENDRON ARALIOIDES

I grow this shrub for its glossy ivy-like leaves. This form was originally collected in Taiwan, China by Crûg Farm Nurseries and it grows faster with glossier leaves than the typical Japanese-origin plants. It is a fantastic evergreen shrub.

VIBURNUM BETULIFOLIUM ▼

In 2003, I collected seeds of this in China and although it is a messy business to clean the seeds, the plants do grow well in Ireland. I have found that they produce massive amounts of glossy red berries when you grow them in groups. For long winter interest, this is the best and most reliable shrub for berries at Hunting Brook.

ZANTHOXYLUM AILANTHOIDES

I have a real fondness for pinnate leaf structure, partly because it is useful in gardens to allow light through to the plants below. Like a lot of the good plants at Hunting Brook it has spiny leaves and stems.

ZANTHOXYLUM LAETUM ▼

I got this at a plant fair in Cornwall. It is growing into a small tree in the Valley with extremely spiky leaves and stems. It had red flowers on the stems during late summer and autumn.

FOLIAGE PLANTS

BORINDA ALBOCEREA
I am building a collection of bamboos and have an interest in *Borinda*. This stunning species has a white waxy bloom on the new culms and fine-textured foliage. Native to the mountainous Yunnan province in China, it has an open clumping habit with large blue-green leaves. It can grow up to 3m (10ft). I have planted it on the banks in the Valley.

BORINDA NUJIANGENSIS ▼
A lovely elegant variety with the slenderest leaves of any *Borinda* I grow. The new shoots have beautiful reddish, marbled sheaths and light blue culms growing up to 3m (10ft). It is absolutely gorgeous, and I encourage you to go and find it.

CANNOMOIS GRANDIS
This is a stunning large restio with stout culms, 2m (6½ft) tall. They have a bright reddish tint to them. I grew lots of restios before the winter of 2010 and this was the only one that survived the -15°C (5°F) that winter. I have planted it in full sun and it is extremely well drained. These plants need acid soil.

CAREX OSHIMENSIS
One of the best evergreen grasses for sun or shade, it grows in my woodland beds where it gives great structure during the winter months. If an evergreen grass becomes unkempt, cut it back every second or third spring.

CHIMONOBAMBUSA TUMIDISSINODA ▼
A wonderful arching bamboo which spreads fast. During the expedition in China, we used walking sticks made from it when trekking in the mountains of Emei Shan. I recommend growing it in a large pot as it is a real runner.

CHIONOCHLOA RUBRA
I know I use the word 'favourite' a lot, but this is my favourite grass. I was delighted to find hillsides of it in New Zealand with sheep grazing through it. It looks great all year around and I never do any maintenance on it. My friend Seamus O'Donnell from Cluain na dTor Seaside Nursery and Gardens (see Plant Nurseries), sells this plant – it can be tricky to source.

CHUSQUEA GIGANTEA
The most spectacular bamboo at Hunting Brook, this produces culms up to 6m (20ft) long and is getting thicker each year. If I were asked to recommend a tall specimen bamboo, this would be it.

CORTADERIA RICHARDII SYN. *AUSTRODERIA RICHARDII*
I witnessed this elegant evergreen grass growing wild in New Zealand. At Hunting Brook, I grow it on a bank with *Aralia echinocaulis* to great effect. The flowering stems need to be cut back in the spring.

ENSETE VENTRICOSUM 'MAURELII' ▼
The huge red leaves of this banana make it the best exotic foliage plant in my tropical borders. It needs winter protection and lots of manure when planting, then feed through the growing season. Propagate by offsets in spring or buy seeds from jungleseeds.co.uk and sow in heat during the spring. The leaves on this variety tend not to get damaged in the wind as is the case with other varieties.

FARGESIA FERAX SYN. *F. ANGUSTISSIMA*
I particularly like fine-textured bamboos, and this is one of the most graceful. It is a large, non-invasive, clump-forming bamboo with a narrow base, ideal for a garden where space is limited but you want height. The new culms are a pale misty blue as they develop.

FARGESIA NITIDA 'EISENACH'

A clump-forming bamboo ideal for screening due to its upright growth habit. It is perfect for a small garden as it does not spread. We thin out half the culms every year and then strip one-third of the side stems from the remaining culms to achieve a graceful, see-through appearance.

MISCANTHUS SACCHARIFLORUS

This is the tallest miscanthus and similar to *M. giganteus*, which is used as a bio fuel. It looks similar to sugar cane and provides a tropical effect. It needs light all around as it tends to flop over. It propagates easily from division in spring but tends not to flower in this part of the world.

MUSA SIKKIMENSIS 'BENGAL TIGER' ▼

I bought this from Tony Avent at Plant Delights Nursery in the US. In my opinion, Tony has the most comprehensive private plant collection in the world. This banana has turned into one of the stars of the exotic show at Hunting Brook with its huge leaves and maroon streaks.

RESTIO QUADRATUS ▼

A clump-forming architectural plant from the Fynbos area of South Africa, this likes an open, sunny position with good air movement and well-drained neutral to acid soil. Restios look very different as young plants, having a juvenile growth form of fluffy green, fairly lax stems. In the third year, they start to send up more mature culms, turning into their elegant adult selves. This is a very hardy plant, surviving up to -15°C (5°F) at Hunting Brook. It has 1.5m (5ft) high bushy horse's tails with square stems. I sometimes cut this plant for the house during special occasions.

THAMNOCALAMUS CRASSINODUS 'KEW BEAUTY'

A real gem and one of my favourite bamboos, it has billowing masses of exquisite tiny leaves on upright blue-grey canes. It is very tightly clump-forming, with a height of approximately 4m (13ft), and is hardy to about -15°C (5°F). This bamboo is available in most good garden centres – I was fortunate to get 20 when a nursery was closing down a few years ago.

THAMNOCALAMUS CRASSINODUS 'LANGTANG'

The three selections of *Thamnocalamus crassinodus* that I grow – 'Kew Beauty', 'Merlyn' and 'Langtang' – are all incredible and worth tracking down. 'Langtang' is the most graceful as it forms a froth of weeping foliage with tiny leaves. They all produce pale blue culms and a tight base on the plant that does not spread.

THAMNOCALAMUS CRASSINODUS 'MERLYN' ▼

One of the most elegant and delicate-looking bamboo plants, with very small leaves and distinctive new blue-grey canes. Regular thinning of the canes is essential. This one has come through -15°C (5°F). It is named after Merlyn Edwards, who introduced it to the UK and Ireland.

HERBACEOUS PERENNIALS

ACANTHUS ARBOREUS
Another spectacular spiny plant and yes, it is a tree acanthus. Originally from Yemen, it is probably not hardy at all. It is an evergreen and has pink spiky flowers.

ACONITUM CARMICHAELII 'ROYAL FLUSH'
Bronze foliage appears early in spring and the plant flowers during late summer with deep blue flowers. It is a reliable perennial for flowering in September. Beware – the entire plant is poisonous.

ACTAEA CORDIFOLIA 'BLICKFANG' ▼
A few years ago, I visited Hessenhof Nursery in Holland and one plant that stood out for me in their stock beds was *Actaea cordifolia* 'Blickfang'. It is slow to bulk up but produces wonderful spires and looks great woven through the main border. Christopher Lloyd always said you cannot have enough spikes in the garden. It needs soil that will not dry out.

AEONIUM 'CYCLOPS'
A huge aeonium with dark leaves and a distinct green eye in the centre compared to 'Voodoo'. I buy a lot of my aeoniums from Surreal Succulents mail order in Cornwall. Cyclops is a cross between *Aeonium undulatum* and *Aeonium arboretum* 'Zwartkop'.

AEONIUM 'VOODOO'
I collect aeoniums but have unfortunately lost the names of most. *Aeonium* 'Voodoo' stands out from the rest as it is huge and dark-leaved. I combine these with low perennial planting in Fred's Garden.

AGASTACHE 'BLACK ADDER'
There are so many blue agastache and in my opinion this is the best. Protect against slugs in early spring – slug damage and winter wet are two of the main reasons for agastache not returning in the spring. Propagate by seed in spring or cutting from the new growth.

ALLIUM 'PURPLE RAIN'
Last year was the first time I grew this allium – I planted 500 in the garden meadow. It is supposed to come back reliably each year. It is a cross between *A.* 'Purple Sensation' and *A. cristophii* which gives it a more open structure, longer flowering and better colour holding compared to 'Purple Sensation'.

ALSTROEMERIA 'ORANGE GLORY'
This is a long-flowering alstroemeria with bright orange flowers. It dislikes root disturbance, so plant it where you intend to keep it When deadheading alstroemeria, pull stems that have finished flowering out of the ground to encourage new flowers. www.postalplants.co.uk is a great nursery that specializes in *Alstroemeria*. Always go for the medium-to-large alstroemerias as the small varieties are usually ugly!

AMICIA ZYGOMERIS
This is a woody perennial plant in the family Fabaceae (legumes). Native to Mexico, it is said to be hardy down to -10°C (14°F). Helen Dillon grew this outside through the winters for many years. It has leaves composed of two heart-shaped leaflets and purple bracts which fold down at night and open during the day. The yellow pea-like flowers appear in September. Propagate by cuttings in September.

ANGELICA EDULIS
A perennial umbellifer which grows to 2m (6½ft), with large, white umbels, reddish stems and glossy leaves. This is a wonderful dramatic plant which I am planning to use as repetition planting through one of my borders.

ANGELICA PUBESCENS ▼
I first saw this angelica growing on the terrace in Arabella Lennox-Boyd's garden. While it took some time to settle in, it finally flowered at Hunting Brook during autumn 2018 and still flowers heading into December. The flowering stem grows up to 2.4m (8ft) with a white sparkler-like umbel.

ARTEMISIA STELLERIANA 'BOUGHTON SILVER' ▼

Also known as 'Mori's Form', this plant has been used to edge the new bed at Hunting Brook, where it has formed a dense mat-forming plant. I love the way it creeps into the border and softens the planting.

ARUM ITALICUM SUBSP. *ITALICUM* 'SPOTTED JACK'

I am currently building a collection of arums. This one is an excellent winter/early spring plant with black spotted green, arrow-shaped leaves. I remove the seeds to prevent them spreading. This looks great combined with snowdrops.

ASTELIA FRAGRANS

A great evergreen and totally hardy. Many years ago, I got some from Dr Neil Murray's nursery when it was closing down. I use them repeatedly through the gardens and Valley for winter structure. Division occurs in the spring for any plant with this shaped leaf (parallel veining). I observed these growing wild in New Zealand, including growing on trees.

ASTELIA NERVOSA 'BRONZE GIANT'

This extremely rare golden astelia is beautiful but very slow to bulk up – I wish I had more of it! Originally, I got it from Ballyrogan Nurseries in Northern Ireland. The only other two plants I know of are in Mount Usher Gardens and Kilmacurragh Botanic Gardens (both in Co. Wicklow).

ASTILBE CHINENSIS VAR. *TAQUETII* 'PURPURLANZE' ▼

A late-flowering variety that has tall and elegant spires of magenta flowers from mid-summer to autumn with two months' season of interest. It needs to grow in soil that does not dry out and is easy from division. I am generally not a huge fan of astilbes, especially the small ones, but this possesses a lovely tall and natural form.

ASTRANTIA MAJOR 'BO-ANN' ▼

I sourced this many years ago from the Bali-Hai Nursery in Northern Ireland and it is one of the longest-flowering plants at Hunting Brook, from March until November. I cut the flower stems back (not the leaves) as the flowers fade and it will quickly flower again.

BACCHARIS GENISTELLOIDES

This is possibly the most bonkers-looking plant at Hunting Brook! It has flattened leafless stems and small white flowers at the tips. This plant needs protection from winters in Ireland; it is considered an extremely important medicinal plant in its native South America.

BAPTISIA 'DUTCH CHOCOLATE'

I got this from the Potting Shed Nursery in Wexford, where I find many of my new perennials. It is a real beauty with chocolate-coloured flowers. I am finally achieving success with baptisia after many years of trying; I now have them planted in a south-facing area with extremely good drainage.

BEGONIA TALIENSIS ▼

This is the best in my collection of hardy begonias. It is a recent introduction from China, with dramatic green leaves marbled dark green-black and silver. At the end of the season, it has pale pink flowers which are edible (like all begonia flowers).

BLECHNUM CHILENSE

A big evergreen fern of Chilean origin, this needs moist, acidic soil to spread. To keep the plant looking good, I cut the previous year's fronds back each spring. It is covering a bank near the stream in the Valley.

HERBACEOUS PERENNIALS

BUPLEURUM LONGIFOLIUM 'BRONZE BEAUTY'

I am weaving this plant through the border. It has large yellowish-green flowerheads which turn coppery-bronze in autumn and is a beautiful cut flower.

CALANTHE TRICARINATA ▼

A hardy orchid for the shade garden, this is now well established at Hunting Brook since 2010 and thriving in the Woodland Garden, where I have divided it and replanted it into very good leafmould.

CANNA 'BIRD OF PARADISE' ▼

A magnificent canna which I obtained from Urban Jungle in Norfolk, this has never actually flowered but has the most wonderful foliage similar to a bird of paradise leaf. It grew to 2.1m (7ft) high this summer.

CANNA 'MUSIFOLIA'

I grow this huge canna in my exotic plantings, where it reaches 2.4m (8ft), for its spectacular foliage as it has never flowered with me. I provide plenty of manure and extra feed through the summer.

CANNA 'TANEY'

When I moved to Hunting Brook, I presumed it would be too cold to grow cannas for flowering due to the lack of summer heat. After a plant-shopping trip to Urban Jungle, I came home with a collection of cannas which grew successfully for their foliage. I chose tall varieties and 'Taney' was one of the best, with glaucous leaves growing up to 2.4m (8ft) and flower stems reaching 3m (10ft), with apricot-coloured blooms.

CARDAMINE ENNEAPHYLLA ▼

Dan Hinkley has had a huge influence on my gardening life through his books, lectures and his garden, Windcliff. His book *The Explorer's Garden* was where I first saw cardamines discussed. *C. enneaphylla*, a gift from Helen Dillon, is the earliest-flowering shade plant in my Woodland Garden, with soft yellow blooms. It is slow to bulk up but totally unaffected by frost and flowers no matter the weather. I often wonder why it is not grown more.

CARDAMINE HEPTAPHYLLA

Seven-leaved toothwort flowers really well every year; however, it is prone to spreading, so keep an eye on it. It is a beautiful plant with white-lilac flowers like sweet rocket in March/April.

CARDIMINE PENTAPHYLLA

A reliable spring-flowering perennial for shade with drooping purple flowers, this is the best of my cardamine collection because it doesn't spread and the flower colour is stronger than that of other cardamines. Keep an eye out for this in the spring plant fairs.

CARDAMINE QUINQUEFOLIA

This small woodland plant with attractive lobed leaves and violet-mauve flowers spreads moderately at Hunting Brook. It could be considered invasive but I love it and it dies back not long after flowering.

CELMISIA SEMICORDATA 'DAVID SHACKLETON' ▼

I am very lucky to have this incredible silver celmisia which grows well and propagates at Hunting Brook. I was given it by Beach Park Gardens near Dublin many years ago. I saw celmisia growing in Arthur's Pass, New Zealand, which was a damp, cool area even in summer, similar to Hunting Brook.

CHRYSANTHEMUM 'RUBY MOUND' ▼

On a recent trip to the USA I was astounded by the variety of hardy chrysanths flowering in gardens. They are so useful for extending the season and I am trying out a selection at Hunting Brook to see how they will perform when we do not have a warm September. 'Ruby Mound', with deep maroon flowers, is thriving in the new border.

CHRYSOSPLENIUM DAVIDIANUM

This plant makes good groundcover, forming a thick carpet over the soil. You need to have the perfect conditions of damp shade, which I do in the Valley, otherwise it looks unsightly and is not worth growing.

CHRYSOSPLENIUM MACROPHYLLUM

Discovered as recently as 1996 on Emei Shan in China, this is a must-have plant for its distinctive bergenia-like foliage and, in early spring, large umbels of white-tinged green flowers, borne in the middle of the softly hairy leaves. Once established, it will send out long runners to extend its territory and form sizable colonies, if given adequate moisture and shade. It is extremely happy in the deep shade and damp of Hunting Brook's Valley.

CIRSIUM LANATUM

This plant was collected by Seamus O'Brien in Tibet, China a few years ago. It is a giant thistle, growing up to 3.6m (12ft), with vicious, spiky leaves and drooping yellow thistle-type flowers. It is a real showstopper in the Valley.

CORYDALIS 'BLACKBERRY WINE'

This is a hybrid corydalis with fern-like blue-green leaves and fragrant, wine-purple flowers from May to July. If the foliage is cut back after blooming, there may be a sparse reblooming later in the summer or early autumn. It likes to be cool in summer and in soil that does not dry out; given those conditions, plants may even bloom throughout the summer. I am slowly building up a collection of corydalis as much for their foliage as their flowers.

CORYDALIS 'CALYCOSA' ▼

There are so many blue corydalis now and I reckon this is the best; it grows up to 45cm (18in) and flowers for two months, as long as it does not dry out. I bought it from Peter Korn at the Great Dixter Plant Fair, East Sussex.

CORYDALIS SOLIDA 'FIRE BIRD' ▼

C. solida is great for combining with snowdrops, as long as you keep an eye out for the latter taking over. 'Fire Bird' is the best of the red solidas, the perfumed bright, rich red flowers are held above ferny, divided foliage.

CORYDALIS TEMULIFOLIA 'CHOCOLATE STARS'

I grow this for its amazing brown spring foliage which appears in February; the flowers are a nondescript, dirty blue. It is an easy plant to grow and I combine it with snowdrops to great effect.

COSMOS PEUCEDANIFOLIUS

This perennial pink cosmos has been growing outside at Hunting Brook for 10 years without any protection in winter; it is sited on top of a raised wall where it has excellent drainage and full sun. Like the chocolate cosmos *C. atrosanguineus*, it forms a tuber. It should be grown more in gardens.

COSMOS SULPHUREUS 'TANGO'

I took a tour to RHS Wisley in Surrey when a *Cosmos* trial was on in the trial grounds. This one was the winner for me with its vibrant orange flowers. Orange cosmos do not usually flower well at Hunting Brook as they need a warm summer, but 'Tango' starts flowering early and goes right through to late autumn. Thanks to Thomas at Mr Middleton Garden Shop in Dublin for sourcing these seeds for me.

HERBACEOUS PERENNIALS

DAHLIA AUSTRALIS
I obtained this plant at Crûg Farm Nurseries a few years ago and now have huge tubers with 1.8m (6ft) stems covered in single pink flowers. This species starts flowering in July, earlier than many species dahlias. I saw hillsides of similar dahlias in Costa Rica.

DAHLIA 'BRIGHT EYES'
When it comes to growing dahlias I mainly use my own seedlings, though I still resort to a few purchased ones and this single-flowered plant is one of them. It really stands out with its fuchsia pink and yellow centre.

DAHLIA 'KILBURN GLOW'
This waterlily-type dahlia has luminous bright fuchsia pink shading to yellow near the base of the petals.

DAHLIA 'NIGHT BUTTERFLY' ▼
'Night Butterfly' is a real showstopper with its velvety deep red outer petals and inner layer of pink and white. Bees love it.

DAUCUS CAROTA 'PURPLE KISSES'
A gorgeous umbellifer which I grew for the first time last spring from seeds supplied online by Sarah Raven. I planted it in combination with *Inulanthera calva* and orange species dahlias. It has a mix of crimson and white in the flowers, which are very long-lasting in the house as a cut flower.

DICENTRA FORMOSA 'BACCHANAL'
I absolutely love this dicentra with its lacy blue-green leaves and rich red flowers. It is an easy woodland plant in soil that does not dry out. If the soil gets very dry during the summer, it will go dormant.

DRYOPTERIS WALLICHIANA ▼
An outstanding fern with upright stems covered in black-brown bristly hairs in spring. It can grow in dry, shady conditions once established. I remember Christopher Lloyd of Great Dixter saying that it was his favourite fern.

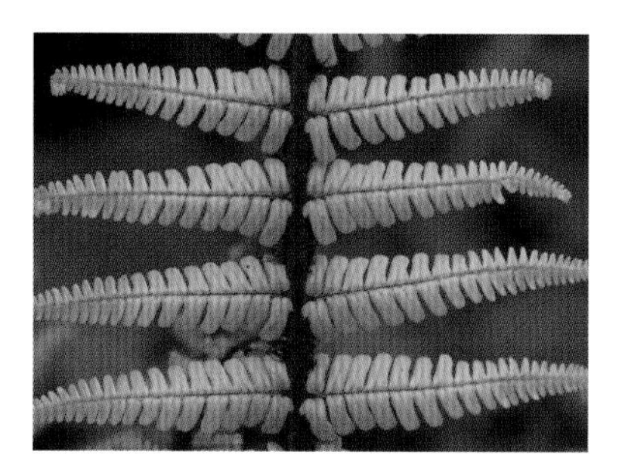

EPIMEDIUM 'AMBER QUEEN'
One of the best new epimediums; what I particularly like is that the flower stems are well above the foliage. Heart-shaped leaves with burgundy mottling emerge in spring, along with long flowering sprays of amber flowers. It can tolerate drought and root pressure, forming clumps or spreads and creating a superb ground cover.

EPIMEDIUM 'DOMAINE DE SAINT JEAN DE BEAUREGARD'
This is a very nice selection from Thierry Delabroye, a French nurseryman. I purchased it from him at the Chantilly Plant Fair. It is a long name for a plant with such tiny flowers – they are a beautiful yellow and pink and very plentiful. The foliage is attractively mottled.

EPIMEDIUM 'DOMINO' ▼
Of all the epimediums I grow, this is my favourite. It has beautiful spring foliage – green leaves splashed with bronze – and white spurred flowers with maroon tips over a long period of time.

EPIMEDIUM MEMBRANACEUM
This has lovely mottled leaves in spring and yellow flowers from May until winter – one of the longest-flowering epimediums we have at Hunting Brook. What a hard-working epimedium!

EPIMEDIUM 'ROYAL PURPLE'
'Royal Purple' has purple/bronze new growth in spring, followed by lavender-blue flowers. This one is extremely choice and worth tracking down.

EPIMEDIUM 'SPINE TINGLER'
This is another new epimedium and what a gem it is. It has lance-shaped, toothed leaves, bronze when young, and sprays of yellow flowers in spring. Epimedium foliage is very important to me and this certainly passes the test. I cut back old leaves in late winter or early spring before flower spikes form.

EPIMEDIUM 'THE GIANT'

First discovered in 2001 by epimedium guru Darrell Probst near Chongqing in China, where I also did some plant hunting, this has large leaves and long stems up to 1m (3¼ft) culminating in a spray of spider-like yellow and amber-centred flowers. I was wildly excited when I saw that it grows to 1–1.8m (3¼–6ft), although it is unfortunately not quite that size at Hunting Brook. However, it is still worth growing.

EPIMEDIUM 'WILDSIDE AMBER' ▼

An introduction by Keith Wiley, this has beautiful young foliage and apricot flowers. It is a real gem, worth keeping an eye out for.

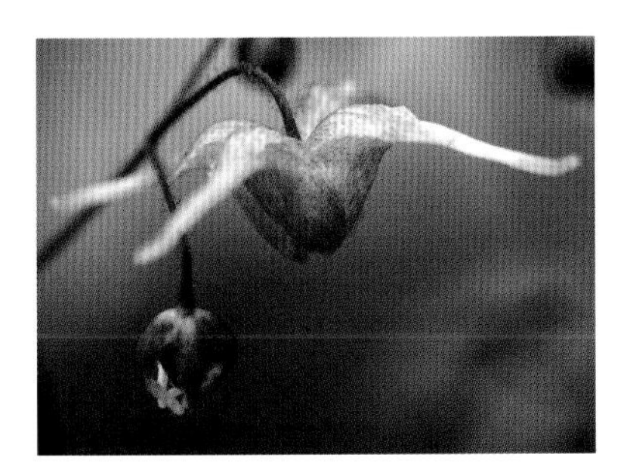

EPIMEDIUM 'WILLIAM STEARN' ▼

I have grown this for many years and it has always been one of the best in my collection of epimediums. It has deep red flowers above the foliage. I like to plant them on a slope in the Valley to help visitors appreciate the dainty hanging flowers.

EPIMEDIUM WUSHANENSE 'CARAMEL'

An epimedium with a difference, having evergreen, thick and narrow holly-like leathery foliage that is red and mottled during spring. The caramel-coloured flowers are held in large numbers well above the foliage on tall racemes.

ERANTHIS HYEMALIS 'ORANGE GLOW'

This has a bright orange flower, subtly distinct from the normal acid lemon flowers of this species. It was derived originally from Copenhagen Botanic Garden and usually comes true from seed. It is like a light bulb in my Woodland Garden in January and February. Buying them in spring with their leaves is better than purchasing dried-up bulbs in autumn.

ERANTHIS HYEMALIS 'SCHWEFELGLANZ'

A vigorous clone with large, straw/apricot-coloured flowers over dark green foliage. The name means 'sulphur glow' and any plant that flowers in late winter, especially in a cheerful colour, is worth collecting.

ERYNGIUM DEPPEANUM

This was a new plant for me last year with incredibly spiny leaves and the appearance of an agave. It is a great addition to Fred's Gardens. The flowering stem has silvery-green spiky flowers.

ERYNGIUM EBURNEUM ▼

I have grown this eryngium for 20 years for the best and most reliable seedheads through the winter and spring. It is sited on a sunny bank where its rosettes of spiky leaves look very exotic (in reality, the plant is totally hardy). A South American species, it seems to thrive at Hunting Brook.

ERYNGIUM GUATEMALENSE

This eryngium was collected by Sue and Bleddyn of Crûg Farm Nurseries in 2014 in the Plateau al Sierra del Guatemala. The stems bear black-brown thimbles surrounded by silver-blue pointed bracts. I have planted a group of them in Fred's Garden for their spiky effect.

ERYNGIUM PANDANIFOLIUM 'PHYSIC PURPLE'

An architectural plant that is the tallest-flowering eryngium, reaching 3m (10ft). It has long, sword-like, silvery glaucous green leaves and sprays of dark red to purple flowerheads. It requires full sun and soil that remains moist. This is the best form with purple flowers and was still in flower in December this year.

HERBACEOUS PERENNIALS

ERYTHRONIUM 'HARVINGTON SNOWGOOSE' ▼

My favourite erythronium, it has big white flowers with the most elegant of petals. It is a real showstopper that bulks up quickly. My friend Finlay Colley always has a wonderful selection of erythroniums at the spring plant fairs in Ireland.

ERYTHRONIUM HENDERSONII

This erythronium has excellent foliage with lavender flowers and wonderful dark anthers. You cannot get a plant more elegant than this in a spring garden with its gorgeous flowers that bulk up very fast for me. It blooms earlier than most.

ERYTHRONIUM 'JOANNA' ▼

This is one of the most beautiful of all my erythroniums, with pale creamy-yellow flowers turning apricot as they age on tall stems. This seems to be rare in cultivation, so if you ever see it for sale, grab it!

ERYTHRONIUM 'MINNEHAHA' ▼

A most elegant erythronium, this has white flowers and reflexed petals. It was a present from my friend John Grimshaw a few years ago. Never buy in autumn when the bulbs are usually dead – instead, buy them in spring as potted plants with leaves.

ERYTHRONIUM REVOLUTUM 'KNIGHTSHAYES PINK'

This is one of the pinkest-flowered American species with mottled foliage and gorgeous rosy blooms with prominent yellow anthers. 'Knightshayes Pink' was developed at the National Trust garden of that name in Devon and it is bulking up quickly for me.

ERYTHRONIUM TUOLUMNENSE 'SPINDLESTONE' ▼

A variety with good early yellow flowers and quick to bulk. I divide these in May, just as they start to die back.

EUPATORIUM CHINENSE

In 2004, I collected the seeds of this plant in China. It is a much more refined eupatorium than the usual tall ones. The leaves are narrow and have a glaucous look. It is a wonderful perennial that should be more readily available to purchase.

GALANTHUS NIVALIS 'GREEN TEAR' ▼

A very special present from John Grimshaw a few years ago, it is slowly starting to bulk up into a nice clump.

GALANTHUS PLICATUM 'TRYMLET' ▼

I generally find November and December the most challenging months as a gardener, but over the years my ex-partner Trevor always helped me to connect back with my plants by encouraging me to go out and watch the cycle of life starting again in the garden with the snowdrops peeping up and starting to bloom. For me, collecting snowdrops bridges the gap from the depths of winter into the growing season starting again and reminds me of how much I love what I do. This is the fastest snowdrop with green markings to bulk up. It was worth the large amount of money I spent on it a few years ago.

GALANTHUS 'SPINDLESTONE SURPRISE' ▼

I have tried a few yellow snowdrops and 'Spindlestone Surprise' is the one that bulks up quickly. This was a present from my good friend Anna Nolan, who started me off growing snowdrops and so many other plants. Sadly Anna has passed away, but I will never forget her generosity and her plants still thrive at Hunting Brook.

GALANTHUS 'TRUMPS'

In early February I always attend the Snowdrop Gala and join in the frenzy of snowdrop-shopping with all the other gardeners coming out of hibernation after winter. I always get a few snowdrops to add to my collection – I am fond of all of them and especially those with green markings like this beauty.

GERANIUM 'ANNE THOMSON'

This was one of the most iconic plants at Hunting Brook; I had it planted along the length of the main bed for 14 years. It produces deep magenta flowers with a black eye from early June until October. There is no need to cut it back in summer, but it must have full sun and good drainage. Propagate by division, even during the summer.

GERANIUM PRATENSE 'SOUTHEASE CELESTIAL'

Plant in full sun and you will enjoy large powder-blue flowers in May and June. I bought this from Graham Gough at Marchants Hardy Plants, East Sussex, which is one of my favourite perennial nurseries in the UK.

GERANIUM PSILOSTEMON 'MOUNT VENUS' ▼

This geranium is a seedling from my local perennial nursery, Mount Venus, where I obtain a lot of my perennials. It has a larger magenta flower with a black eye than the usual *Geranium psilostemon*. I use it for edging borders and mass planting on the steep bank as you enter the garden.

GERANIUM WALLICHIANUM 'CRYSTAL LAKE' ▼

Another good geranium that will softly weave through other plants, bearing blue flowers with purple veins and white eyes. It is sterile, so propagation is by division.

GERANIUM WALLICHIANUM 'HAVANA BLUES'

This cultivar has large lavender-blue flowers with dark veins and new yellow foliage in spring. I first fell in love with it when I saw it growing at Dove Cottage Nursery and Garden in Halifax. There are many new *G. wallichianum* cultivars on the market and all are worth growing.

GEUM 'FARMER JOHN CROSS'

I love this geum with its nodding yellow flowers over a period of six weeks. Division of Geums every few years is important to keep them healthy and flowering.

GEUM 'KARLSKAER'

This geum produces warm apricot flowers for about three weeks, so it is not long-flowering but worth it for that colour. I have it planted on the raised bed at the entrance where it is easy to see the small plant.

HERBACEOUS PERENNIALS

GEUM 'SCARLET TEMPEST' ▼

I rely a lot on geums for bridging the gap from spring into summer. In Fred's Garden, I planted more than 100 of them for this reason, using *G.* 'Scarlet Tempest' and 'Totally Tangerine'. The former is not as long-flowering as the latter but provides a splash of colour.

GEUM 'TOTALLY TANGERINE'

I went through a phase of collecting geums, grabbing any I could get my hands on. However, *G.* 'Totally Tangerine' is by far the best and longest-flowering, from April through to winter. It is a sterile plant, so there is no need for dead heading unless it looks untidy.

HEDYCHIUM FORRESTII

This is by far the hardiest and most reliable hedychium to flower in my garden. They are late risers and do not appear above ground until early summer – in fact, many a time I have planted a dahlia on top of them! The foliage provides an exotic appearance and the white flowers are useful for extending the season into September.

HEDYCHIUM WARDII ▼

I am planting up a hillside in the Valley with a collection of hedychiums; however, some may not be hardy. This recent addition has spectacular late-flowering bright yellow flowers. It is easy to propagate from division in spring.

IMPATIENS OMEIANA 'SANGO' ▼

I got this from Nick Macer at Pan-Global Plants. The leaves have a band of metallic pink and a silver dusting along the veins and it bears large yellow orchid-like flowers in the autumn. It is a spreader, so plant with caution.

INULANTHERA CALVA

This is the quirkiest, funkiest foliage plant I grow. It produces woody shoots like bottle brushes up to 90cm (3ft) long, which I have interplanted through the new border to give it a bonkers look. The label said it would flower in December, which means I shall never see flowers on it.

LATHYRUS AUREUS

This a superb dwarf, clump-forming perennial pea, cousin to the familiar sweet pea. It forms a bushy mound of light green lacy leaves with spikes of orange pea flowers appearing in early summer.

LEUCOJUM VERNUM VAR. *CARPATHICUM*

An absolute beauty for the spring garden with its yellow-rimmed white flowers. It requires the same growing conditions as snowdrops. There are interesting new cultivars available, but I have not managed to get my hands on those yet!

LIGULARIA 'BRITT MARIE CRAWFORD'

A plant with dark wine-coloured leaves and orange flowers, easy to divide in early spring. I have this growing in a damp bed at the side of the house. The bed has a succession of interest with snowdrops followed by *Narcissus* 'Jetfire', which looks great with the wine-coloured foliage of the ligularia.

LINARIA 'PEACHY'

This is one of the repeat plants through Fred's Garden. I love its mix of peachy/apricot flowers which bloom non-stop from early summer until well into autumn. It is the very best flower for bumble bees at Hunting Brook. Thank you to Paul and Orla Woods at the fantastic Kilmurry Nursery, Co. Wexford, for introducing me to this plant.

LINARIA 'ONSLOW PINK' ▼
Of all the new linarias I trialled this year, this was the longest-flowering and had beautiful flowers of pink and white. Most of the new linarias came from Cotswold Garden Flowers.

LINARIA TRIORNITHOPHORA
The flowers of this species are purple or pink and arranged around a stem in groups resembling budgerigars – in fact, the common name for this plant is 'three bird toadflax'. It needs sun and well-drained soil.

LINARIA VULGARIS F. *PELORIA* ▼
A quirky linaria with lemon-yellow flowers, it is unlike any other toadflax in that the flower forms a rounded tube with five spurs on willowy stems of grey foliage. It spreads gently in the border and through my dry stone walls.

LOBELIA × *SPECIOSA* 'HADSPEN PURPLE'
I have trialled many lobelias at Hunting Brook and 'Hadspen Purple' is the most reliable. I weave them through Ashley's Garden. Lobelia need damp soil and you must watch the slugs, but they are easy to propagate from division in March. During the first few years, I dig them up for winter and keep them potted in the tunnel, then plant out again in May.

LOPHOSORIA QUADRIPINNATA
This hardy evergreen fern from Chile has flourished on the banks in the Valley at Hunting Brook. The magnificent fronds (leaves) have a silvery blue underside and the fern can grow up to 3m (10ft) in height. It needs to become more available.

LYCHNIS CORONARIA 'GARDENERS WORLD' ▼
A short-lived perennial with double purple/magenta flowers, this needs well-drained soil in full sun. It is sterile, so self-seeding is not a problem. It was a favourite of Monty Don's when he came to visit Hunting Brook. I propagate it by basal cuttings in spring.

LYCHNIS 'HILL GROUNDS'
With silver leaves and bright pink flowers, this is an easy short-lived perennial for full sun and well-drained soil. It is a sterile hybrid between *L. flos-jovis* and *L. coronaria.* It has the most vivid pink flowers at Hunting Brook - to think I did not really use much pink until recent years! It is ideal for flowering in early summer combined with *L. coronaria* 'Gardeners World' and is propagated by basal cuttings in spring.

LYSIMACHIA BARYSTACHYS 'HUNTINGBROOK'
I collected the seed in China and when it germinated, I selected the one seedling with red stems rather than the usual green stems. The flowers are slender, arching white spires which are longer than the more common *Lysimachia clethroides.* It makes an excellent cut flower in the tea room.

LYSIMACHIA PARIDIFORMIS VAR. *STENOPHYLLA* ▼
I absolutely love this late-flowering plant, which is unlike any other lysimachia in that it is well-behaved and it's a woodlander. It has yellow flowers in a rosette rising to 60cm (2ft). I cut back the old leaves in January, as you would do with *Helleborus.*

HERBACEOUS PERENNIALS

LYTHRUM SALICARIA 'LADY SACKVILLE'

I grow many different lythrums and this is the best. It is a good reliable perennial for the late summer and early autumn but does require damp soil to produce those tall pink spires. It would look great in a damp meadow.

MAIANTHEMUM RACEMOSUM SYN. *SMILACINA RACEMOSA*

This is in my top five plants for shade. It has good healthy foliage and flowers that smell of almonds, plus it is easy to grow and propagate by division in autumn. I have this repeated through the Woodland Garden.

MECONOPSIS PANICULATA

With its golden hairy leaves, this would be worth growing as a fantastic foliage plant even if it never flowered. I had it in too much shade last year and it flowered with a skinny bloom falling towards the light. Surface sow seeds fresh in autumn.

MONARDA 'ON PARADE'

A magenta-flowering monarda that blooms over a good few weeks and does not get mildew. It performs best on heavy clay soil. A few years ago, I did a trial of monardas and this was the winner. The best mildew-resistant monardas in my trial were 'Balance', 'Purple Rooster', 'Scorpion', 'On Parade', 'Jacob Klein' and 'Gardenview Scarlet'.

NICOTIANA 'BABYBELLA' ▼

In general, nicotianas do not perform very well for me as Hunting Brook summers are not warm enough. This one was a real surprise to me last summer as it nearly flowered itself to death with deep red flowers from early summer to autumn.

OENOTHERA FRUTICOSA SUBSP. *GLAUCA 'SONNENWENDE'* ▼

This plant possesses the most vivid yellow of any flower. A German cultivar, I got it from Mount Venus Nursery to fill some gaps when Hunting Brook was due to feature on the television programme *Gardeners' World*. I combined it with my black aeoniums to great effect.

PERSICARIA DIVARICATA

Friends brought this back from Finland for me; it is very rare in cultivation. It is willow-like with red stems and a spray of white flowers. Propagate by division.

PODOPHYLLUM AURANTIOCAULE SUBSP. *AURANTIOCAULE* ▼

This was recently introduced from Tibet, China by Ken Cox of Glendoick Gardens Nursery in Scotland. I grow it in the Woodland Garden near the house, where it produces umbrella-like leaves in spring and clear white flowers not hidden by the foliage.

PODOPHYLLUM 'BIG LEAF' ▼

A dramatic foliage plant for semi-shade, I obtained this from Thierry Delabroye Nursery in France. It has huge leaves which appear in March. Interestingly, the deep red/black flowers under the leaves smell like a dead animal, but don't let that put you off the buying the plant!

PODOPHYLLUM CHENGII 'HUNAN' ▼

What an amazing leaf this has, with its strange markings of black, brown and green. It has maroon flowers. I was thrilled to find it in a nursery in Japan, but not many people visiting the garden last summer could quite see why I was so excited by one strange leaf.

PODOPHYLLUM DELAVAYI

This is one of my favourite woodland plants, with mottled brown/green leaves and red flowers. It likes rich, moist soil – avoid direct sun. I find that if you transplant a podophyllum to a different spot, new plants will appear in the original area from root cuttings.

PODOPHYLLUM 'RED PANDA'

I remember seeing this for the first time when I was speaking at a woodland plant conference in Seattle. I dreamt about it for many years until I finally got my hands on it this year. How wonderful it looks in spring, with its copper new foliage turning green in late summer.

POLYGONATUM FALCATUM SILVER-STRIPED ▼

A very good polygonatum with a silver stripe through the leaf and typical white flowers. I am building up a collection of these.

POLYGONATUM VIETNAMICUM

I first saw this growing in Tom Hudson's garden Tregrehran, in Cornwall. I admired it a lot and Tom kindly gave me a plant which is now settling in nicely in the Valley. This Solomon's seal has red flowers and grows up to 1.5m (5ft). Doesn't that sound exciting!

PRIMULA ELATIOR

Native and now extremely rare in the wild, this really is a beautiful soft yellow primula for flowering in May. I have this repeated in Ashley's Garden and cut the leaves back after flowering just to tidy it up.

PRIMULA 'HALL BARN BLUE' ▼

Another group of plants I collect! This is a good blue with a yellow eye and one of my favourite primulas for the spring garden. Check out Barnhaven Primulas.

PRIMULA 'JUNE BLAKE' ▼

My sister, June Blake, sowed seeds of cowslips and this seedling appeared in the middle of the group. It has vibrant yellow flowers in spring and autumn and sporadically throughout the year. I have planted it through Ashley's Garden at Hunting Brook.

PULMONARIA 'BLAKE'S SILVER'

June found this pulmonaria growing in our mother's old garden – it stood out as it has pure silver foliage and beautiful blue/pink flowers. An excellent pulmonaria, it is now available for sale around Europe and America.

PULMONARIA RUBRA 'REDSTART'

One of the first lungworts to flower, starting in December in Hunting Brook, it has coral blooms above lush apple-green foliage. It is a good source of early nectar for bees early in the year. Cut back leaves when flowers fade.

HERBACEOUS PERENNIALS

ROSCOEA 'HARVINGTON RAW SILK' ▼
Thankfully, we have some great nurseries not far from Hunting Brook. Rare Plants Ireland is owned by Finlay Colley and it has got the absolute best plants for sale. He had a fine selection of roscoeas last spring, so of course I had to buy them all! This one possesses large cream flowers with lots of vigour.

ROSCOEA HUMEANA 'SNOWY OWL' ▼
I became a *Roscoea* collector last spring! I blame my great gardening friend, Hester Forde, for this – she is an obsessive collector like me. This one has large, pure white flowers.

SALVIA 'AMISTAD'
The longest-flowering salvia at Hunting Brook, it produces deep purple flowers for at least five months. I left it out in winter and it survived. I cut it back to the ground in spring.

SALVIA BLEPHAROPHYLLA 'PAINTED LADY'
Dazzling red flowers stand out beautifully against the attractive glossy foliage and dark stems. It has a spreading habit and flowers from June to November. This small salvia is a must if you like bright red.

SALVIA BUCHANANII
This salvia was presumed extinct until it was found in a garden in Mexico City. It is rare in cultivation and should be grown more as it is a great garden plant. It has large velvety pink flowers on stems about 60cm (2ft) high.

SALVIA CURVIFLORA
Over a number of years, I have trialled tall salvias that can be combined in Ashley's Garden with tall perennials and exotic plants. This is one of the winners. It has a good upright habit with velvety magenta flowers through summer and up to the first frost. I bring this into the tunnels for the winter.

SALVIA DOMBEYI
From August to October, this possesses the largest flowers in the genus with its 9cm (3½in) pendulous flowers. The flowers are vibrant scarlet with nearly black calyxes. I support it as it grows – it can be vine-like. Join the queue to get cuttings from me!

SALVIA ELEGANS 'TANGERINE'
Totally hardy and easy to grow in full sun, with red flowers all summer. I cut this back to the ground in spring.

SALVIA 'ENVY'
I bought this from Andries and Bob at Foltz Nursery in Holland; it was one of their own seedlings. It grows up to 1.8m (6ft) with deep red velvet flowers. This was new to me this year and is currently one of my favourites.

SALVIA 'FLOWER CHILD'
This is a long-flowering salvia at Hunting Brook, producing its deep pink flowers from May to winter. It is great in a container as it grows to only about 45cm (18in) high.

SALVIA FULGENS
Hardy, with large velvet red flowers all summer and into winter, this salvia needs full sun and well-drained soil. It survives outside at Hunting Brook in winter. It was crowned the winner of my salvia trial in 2017.

SALVIA GUARANITICA 'SUPER TROUPER'
A new blue variety which flowers all summer and into autumn at Hunting Brook. This grows up to at least 2.1m (7ft) and has proved to be fairly hardy. If you do not have space for overwintering these big salvias, I would advise you to take cuttings.

SALVIA INVOLUCRATA 'BETHELLII'
A spectacular salvia, growing up to 1.5m (5ft) with huge pink flowers. This is root hardy, so you could chance leaving it out during winter and cut it back in spring.

SALVIA X JAMENSIS 'EL DURANGO'
I bought this at the plant fair in Chantilly; it is a beautiful salvia with coral/orange flowers.

SALVIA × JAMENSIS 'SHELL DANCER'
An absolute gem of a salvia, with deep pink and hot salmon colours. Remember to plant smaller salvias in full light and avoid cluttering them with other plants.

SALVIA MICROPHYLLA 'CERRO POTOSÍ'
This is the best of the magenta-pink microphyllas, starting to flower in early summer and continuing through to winter. I prune it back by two-thirds in spring.

SALVIA 'MULBERRY JAM'
Another of the salvias I use with the tall perennial planting because it has a very upright habit of up to 1.8m (6ft). It produces bright pink flowers from early summer to late autumn.

SALVIA OPPOSITIFLORA
Another fantastic orange salvia with tubular flowers growing up to 1.5m (5ft). It is tender and where I live in Co. Wicklow it needs to be brought in for winter.

SALVIA PATENS 'GUANAJUATO' ▼
The largest-flowered *S. patens*, it has deep blue flowers and black blotches on the leaves. I am slowly building up stock of this for sale at Hunting Brook.

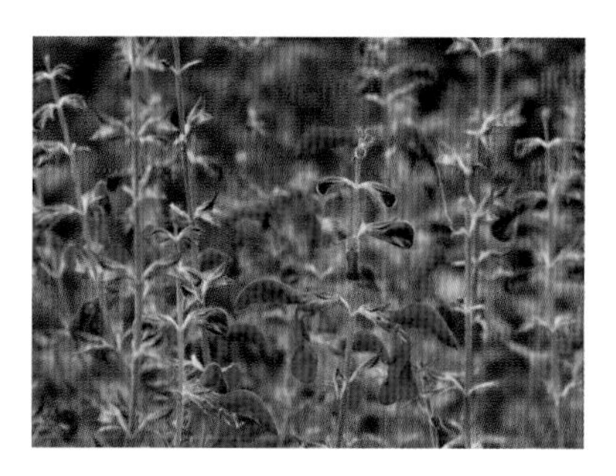

SALVIA STACHYDIFOLIA
This is fast becoming another favourite. It grows up to 2.4m (8ft) with light blue flowers all summer. It has produced a lot of seeds and these will be on sale at Hunting Brook.

SALVIA STOLONIFERA
I am a huge fan of the colour orange in my clothes and love to weave it through my plantings too. I used this in the new border this summer. Discovered in the south of Mexico only a few years ago, it has rusty orange tubular flowers on stems which grow up to 60cm (2ft). I am told it is very hardy and will grow in semi-shade.

SALVIA SUBROTUNDA
One of my favourite annual salvias, I have grown it for the last number of summers. It has fabulous red flowers and grows up to 1.2m (4ft).

SANGUISORBA 'BLACKTHORN'
This is one of my favourite sanguisorbas and the one that won my sanguisorba trial a few years ago when I was in my sanguisorba phase! It has soft pink flowers that stay upright and create a lovely see-through look, wonderful when backlit by the rising or setting sun. It is easy from division and not as bulky as other sanguisorbas as the leaves are lower down, which is important to me.

SANGUISORBA TENUIFOLIA 'PINK ELEPHANT'
One of the tallest sanguisorbas, growing up to 2.7m (9ft), with pink flowers. I stake mine to prevent it falling over.

SCILLA LILIOHYACINTHUS
A very fast-spreading and early-flowering bulb. The broad shiny leaves appear early in the season followed by soft blue flowers. I got this from Mount Usher Gardens many years ago and this one comes with a warning that it will cover a border quickly when it seeds around.

SINOPODOPHYLLUM HEXANDRUM VAR. *EMODI* 'MAJUS' ▼
This is the star of the show in my Woodland Garden. It erupts out of the soil in April with its large umbrella-like leaves mottled with black patches. The stem grows up to about 90cm (3ft), with pink flowers over a few days. This is the best form and worth sourcing.

THALICTRUM DELAVAYI VAR. *DECORUM*
One of my favourite perennials and one of the most asked-about plants in the garden, it grows to about 1.2–1.5ft (4–5ft) with delicate ferny foliage and a spray of purple flowers over two months. It always produces a good crop of seed that I sow in autumn and needs full sun and soil that does not get waterlogged.

TRILLIUM CHLOROPETALUM 'BOB GORDON' ▼
This was a gift from my gardening friend, Billy Moore. He originally obtained seedlings from a great gardener in the Northern Ireland called Bob Gordon; one seedling turned out to be this amazing yellow one. I am blessed with the generosity of my fellow gardeners. This is now settling into its new home in the Woodland Garden at Hunting Brook and blooms in spring with its unique yellow flowers.

TRILLIUM KURABAYASHII
This is the most robust and earliest-flowering trillium in my Woodland Garden. It has deep wine flowers with a twist on the petals and large mottled leaves. Give trilliums your best garden compost and keep them in leaf for as long as possible after flowering.

WOODWARDIA RADICANS
The chain fern is a spectacular evergreen plant with long, dark green, arching fronds up to 2m (6½ft) in length. Plantlets can be pinned down in the soil beside the parent in order to propagate it. This fern prefers damp, acidic soil.

the Giant
(Colchicaceae)
ORNAMENTAL
BERRIES
Disporum
sessile f. macrophyllum BSWJ4316
A robust form of this species which is normally only found in cold areas of Japan, as is so often the case with our other collections from the remote Korean island of Ullungdo. Arising from a brittle running white slender underground rhizome with emerging upright green stems branching in the upper areas. Bearing oblong-elliptic leaves to 16×9.5cm and terminal inflorescences of up to 4 nodding tubular-campanulate white green-tinged flowers to 4cm long April- June. Easily grown in a moist humus rich soil with adequate drainage. in light to full shade
ANEMONE NEMOROSA 'WESTWELL PINK'
I like this more than any other wood anemone. The flowers open white but soon turn the brightest pink. Apr-May, 20cm. Found in Westwell Woods near Ashford Kent. (Ranunculaceae) H7(below -20C)
CORYDALIS curviflora
Choice and rare Chinese corydalis with blue flowers on low creeping stems for moist but well drained, shaded, humus rich site. 15-20cm.
G. Sickle'
Primula Jim saunders
Corydalis Black Berry
Erianthus 'Noel Ayres'
Burgundy knobhead flowers
foliage Jun-Aug. 75cm.
H7(below -20C)
Sorbus harrowiana
Unusual shrubby Sorbus.
Large felty brown
White flowers,
Frangula alnus Asplenifolia
Cut Leafed Alder Buckthorn
A bushy, deciduous, slow-growing shrub with finely textured, scalloped leaves, this
grows to 12 feet tall and almost as wide. 'Asplenifolia' bears clusters of green flowers
followed by round red fruit that ripens to black in the fall. Grow in a shrub border or a
hedging.

Illicium
simonsii
Camellia 'Night Rider'
Others for sale
LINARIA 'TARTE AU CITRON'
crassinodus
the Giant
Epimedium 'William Stern'
Geranium wallichianum Sylvia's Surprise
PLANTS 01452 741641
Pinus bhutanica KR 8654
Harvington Erythroniums
revolutum
'Knightshayes'
Litsea japonica
A small everegreen from Japan. Leaves dark glossy green; flowers yellowish white.
Epimedium sagittatum 'Warlord'
A spectacular foliage plant on account of the striking obliquely arrow-shaped exceptionally large leaflets and especially in the colour of new growth in this form being very heavily speckled in rich maroon and narrower than typical. A rarely seen or available Chinese species with tiny white and yellow flowers but in heads of 50 or more in spring. 45cm. Semi-shade and humus.
Schizachyrium scoparium 'Standing ovation'

FAVOURITE PLANT NURSERIES

Note that opening hours vary and some nurseries offer mail order only. You are advised to view the website for further information before planning a visit.

IRELAND

Altamont Plant Sales
Ballon
Co. Carlow
altamontplants.com

Bali-Hai Nursery
mailorderplants4me.com

Beechill Bulbs
bulbs.ie

Caherhurley Nursery
Caherhurley
Bodyke
Co. Clare
+353 87 9062987

Camolin Potting Shed
Bolinaspick
Camolin
Co. Wexford
camolinpottingshed.com

Cluain na dTor
Ballyconnell
Falcarragh
Co. Donegal
seasideplants.eu

Deelish Garden Centre
Deelish
Skibbereen
Co. Cork
deelish.ie

Future Forests
Maulavanig
Kealkill
Bantry
Co. Cork
futureforests.ie

Green Man Nurseries
greenmannurseries.com

Harrington Exotics
Email: anneharrington6@eircom.net

Johnstown Garden Centre
Junction 8, N7, Dublin Road
Naas
Co. Kildare
johnstowngardencentre.ie

Kells Bay Gardens and Nursery
Kells
Co. Kerry
Kerry
kellsbay.ie

Kilmurry Nursery
Gorey
Co. Wexford
killmurrynursery.com

Leamore Nursery
Cronroe
Ashford
Co. Wicklow
leamorenursery.com

Mount Venus Nursery
Walled Garden
Mutton Lane
Rathfarnham
Dublin 16
mountvenusnursery.com

Mr Middleton Garden Shop
58 Mary Street
Dublin
mrmiddleton.com

O'Driscoll Garden Centre
Mill Road
Thurles
Co. Tipperary
+353 504 21636

Peninsula Primulas
penprimulas.com

Rare Plants Ireland
rareplantsireland.ie

Seedaholic
seedaholic.com

Shady Plants
shadyplants.net

Stam Bamboo
stambamboo.com

Timpany Nurseries
timpanynurseries.com

Two Lions Garden Plants
twolions.org.uk

UK

Ashwood Nurseries
Ashwood Lower Lane
Ashwood
Kingswinford
West Midlands DY6 0AE
ashwoodnurseries.com

Avon Bulbs
Burnt House Farm
Mid Lambrook
South Petherton
Somerset TA13 5HE
avonbulbs.co.uk

Avondale Nursery
avondalenursery.co.uk

Binny Plants
Binny Estate
Ecclesmachan Road
Uphall
Scotland EH52 6NL
binnyplants.com

Cally Gardens
Gatehouse of Fleet
Castle Douglas
Scotland DG7 2DJ
callygardens.co.uk

Cotswold Garden Flowers
Sands Lane
Badsey
Evesham
Worcestershire WR11 7EZ
cgf.net

Crûg Farm Plants
Griffith's Crossing
Caernarfon
Gwynedd LL55 1TU
crug-farm.co.uk

Dove Cottage Nursery and Garden
dovecottagenursery.co.uk

Dysons Salvias
Great Comp Garden
Comp Lane
Platt
Near Sevenoaks
Kent TN15 8QS
dysonsalvias.com

Edrom Nurseries
Coldingham
Eyemouth
Berwickshire
Scotland TD14 5TZ
edrom-nurseries.co.uk

Edulis Nursery
The Walled Garden
Tidmarsh Lane
Pangbourne RG8 8HT
edulis.co.uk

Elizabeth McGregor Nursery
Ellenbank
Tongland road
Kirkcudbright DG6 4UU
Scotland
elizabethmacgregornursery.co.uk

Glendoick
Glencarse
Perthshire
Scotland PH2 7NS
glendoick.com

Growild Nursery
growildnursery.co.uk

Halls Of Heddon
West Heddon Nursery Centre
Heddon on the Wall
Northumberland NE15 0JS
hallsofheddon.com

Hardy exotics
Gilly Lane
Whitecross
Penzance
Cornwall TR20 8BZ
hardyexotics.com

Hardy Tropicals UK
hardytropicals.co.uk

Hart Cannas
hartcanna.co.uk

Knoll Gardens
Stapehill Road
Wimborne
Dorset BH21 7ND
knollgardens.co.uk

Lime Cross Nursery
Herstmonceux
Hailsham
East Sussex BN27 4RS
limecross.co.uk

Marchants Hardy Plants
2 Marchants Cottages
Mill Lane
Laughton
East Sussex BN8 6AJ
marchantshardyplants.co.uk

Monksilver Nursery
monksilvernursery.co.uk

National Dahlia Collection
nationaldahliacollection.co.uk

Pan Global Plants
The Walled Garden
Frampton Court
Frampton-on-Severn GL2 7EX
panglobalplants.com

Pitcairn Alpines
pitcairnalpines.co.uk

Plant World Gardens
St Marychurch Road
Newton Abbot
Devon TQ12 4SE
plant-world-gardens.co.uk

Rare Plants
rareplants.co.uk

Special Plants
Greenways Lane
Cold Ashton
Wilts SN14 8LA
specialplants.net

Summerdale Garden Nursery
Cow Brow
Lupton
Carnforth
Cumbria LA6 1PE
summerdalegardenplants.co.uk

Surreal Succulents
Tremenheere Sculpture Gardens
Guival TR20 8YL
surrealsucculents.co.uk

Urban Jungle
urbanjungle.uk.com

Viv Marsh Postal Plants
postalplants.co.uk

Wildgoose Nursery
The Walled Garden
Lower Millichope
Munslow
Shropshire SY7 9HE
wildgoosenursery.co.uk

BELGIUM

Koen van Poucke
koenvanpoucke.be

Solitair Nursery
solitair.be

FRANCE

Barnhaven Primulas
Keranguiner
22310 Plestin Les Grèves
barnhaven.com

Thierry & Sandrine Delabroye
mytho-fleurs.com

Pépinière Vert Tige
pepinierevert-tige.fr

NETHERLANDS

Bamboekwekerij Kimmei
kimmei.com

Coen Jansen Vaste Planten
coenjansenvasteplanten.nl

De Hessenhof Biologische Kwekerij
hessenhof.nl

Kwekerij De Kleine Plantage
dekleineplantage.nl

Tuingoed Foltz
tuingoedfoltz.nl

JAPAN

Yuzawa Engei
yuzawa-engei.net

NEW ZEALAND

Keith Hammett
drkeithhammett.co.nz

USA

Annie's Annuals
anniesannuals.com

Cistus Nursery
cistus.com

Far Reaches
farreachesfarm.com

Plant Delights Nursery
plantdelights.com

Swan Island Dahlias Nursery
dahlias.com

FAVOURITE GARDENS

Note that opening hours vary and some gardens open only by appointment. You are advised to view the website or email for further information before planning a visit.

IRELAND

Altamont Gardens
Tullow
Co. Carlow
R93 N882
Email: altamontgardens@opw.ie

Blarney Castle
Blarney
Co. Cork
blarneycastle.ie

Caher Bridge Gardens
Formoyle West
Fanore
Co. Clare
Email: caherbridgegarden@gmail.com

Carmel Duignan's Garden
Email: cbduignan@eircom.net

Cluain na dTor
Ballyconnell
Falcarragh
Co. Donegal
seasideplants.eu

Coolcarrigan Gardens
Coill Dubh
Naas
Co. Kildare
Email: rww@coolcarrigan.ie

Cosheen Garden
hesterfordegarden.com

Dhu Varren Garden
Ballyoughtragh South
Milltown
Co. Kerry
dhuvarrengarden.com

Dublin Zoo
Phoenix Park
Dublin 8
dublinzoo.ie

Fruitlawn Garden
Abbeyleix/Shanahoe
Laois
fruitlawn.ie
Email: arthurshackleton9@gmail.com

Glenveagh Castle Gardens
Glenveagh National Park
Church Hill
Letterkenny
Co. Donegal
glenveaghnationalpark.ie

June Blake's Garden
Tinode, Blessington
Co Wicklow
juneblake.ie

Kells Bay House and Gardens
Kells
Co. Kerry
kellsbay.ie

Killruddery Gardens
Southern Cross
Bray
Co. Wicklow
killruddery.com

Knockrose Garden
knockrose.com
Email: ttknockrose@eircom.net

National Botanic Gardens
(Kilmacurragh)
Kilmacurragh
Kilbride
Co. Wicklow
botanicgardens.ie
Email: botanicgardebs@opw.ie

National Botanic Gardens
(Glasnevin)
Glasnevin
Dublin 9
botanicgardens.ie
Email: botanicgardebs@opw.ie

Lismore Castle Gardens
Lismore
Co. Waterford
lismorecastlegardens.com

Mount Congreve
Kilmeaden
Co. Waterford
mountcongreve.com

Mount Stewart
Portaferry Road
Newtownards
Co. Down
nationaltrust.org.uk

Patthana Gardens
Kiltegan
Co. Wicklow
Email: tjmaher100@gmail.com

Rowallane Garden
Saintfield
Co. Down
nationaltrust.org.uk

Russborough House
Blessington
Co Wicklow
russboroughhouse.ie

The Bay Garden
Camolin
Enniscorthy
Co. Wexford
thebaygarden.com

The Dillon Garden
Dun Mhuire
Seafield Avenue
Monkstown
Co. Dublin
dillongarden.com

UK

Benmore Botanic Garden
4 Uig
Benmore
Dunoon
PA23 8QU
rbge.org.uk

Bodnant Garden
Bodnant Road
Tal-y-cafn
Colwyn Bay
LL28 5RE
nationaltrust.org.uk

Bury Court
Bentley
Farnham
Surrey
GU10 5LY
burycourtbarn.com

Cally Gardens
Gatehouse of Fleet
Castle Douglas
DG7 2DJ
callygardens.co.uk

Cambo Gardens
Kingsbarns
St Andrews
KY16 8QD
cambogardens.org.uk

East Ruston Old Vicarage
East Ruston
Norwich
Norfolk
NR12 9HN
e-ruston-oldvicaragegardens.co.uk

Glendurgan Garden
Mawnan Smith
Near Falmouth
Cornwall
TR11 5JZ
nationaltrust.org.uk

Gresgarth Hall
Caton
Lancaster
Lancashire
LA2 9NB
Email: gresgarth@
arabellalennoxboyd.com

Great Dixter
Northiam
Rye
East Sussex
TN31 6PH
greatdixter.co.uk

Levens Hall
Kendal
Cumbria
LA8 0PD
levenshall.co.uk

Lowther Castle
Penrith
Cumbria
CA10 2HH
lowthercastle.org

Marchant's Hardy Plants
2 Marchants Cottages
Mill Lane
Laughton
East Sussex BN8 6AJ
marchantshardyplants.co.uk

RHS Garden Wisley
Woking
Surrey
GU23 6QB
rhs.org.uk

Royal Botanic Gardens Kew
Richmond
Surrey
TW9 3AE
kew.org

The Salutation Garden
Knightrider Street
Sandwich
Kent
CT13 9EW
the-salutation.com

The Yorkshire Arboretum
Castle Howard
York
YO60 7BY
yorkshirearboretum.org

Tregrehan Garden
Par
St Austell
PL24 2SJ
tregrehangarden.uk

Trelissick
Near Feock
Truro
Cornwall
TR3 6QL
nationaltrust.org.uk

Tremenheere Sculpture Garden
Nr Gulval
Penzance
Cornwall
TR20 8YL
tremenheere.co.uk

Wildgoose Gardens
The Walled Garden
Lower Millichope
Munslow
Shropshire SY7 9HE
wildgoosenursery.co.uk

Wild Side Garden & Nursery
Green Lane
Buckland Monachoram
Near Yelverton
Devon
PL20 7NP
wileyatwildside.com

SWEDEN

Eskilsby
peterkornstradgard.se

NEW ZEALAND

Dunedin Botanic Garden
dunedinbotanicgarden.co.nz

Hamilton Gardens
hamiltongardens.co.uz

Hinewai Reserve
hinewai.org.nz

Larnach Castle
larnachcastle.co.nz

Paloma Gardens
paloma.co.nz

Fishermans Bay Gardens
fishermansbay.nz

USA

Bloedel Reserve
bloedelreserve.org

Chanticleer Garden
chanticleergarden.org

Greater Des Moines Botanical Garden
dmbotanicalgarden.com

Heronswood
heronswoodgarden.org

JC Raulston Arboretum
jcra.ncsu.ed

Juniper Level Botanic Gardens
jlbg.org

Ruth Bancroft Garden & Nursery
ruthbancroftgarden.org

The Huntington Botanical Gardens
huntington.org

The Scott Arboretum at Swarthmore College
scottarboretum.org

University of California Botanical Garden at Berkeley
botanicalgarden.berkeley.edu

Windcliff Garden
danieljhinkley.com

INDEX

Page number in *italic* type refer to pictures.

D

E

INDEX

F

G

H

I

J

K

L

M

N

O

P

INDEX

U

V

W

Y

Z

PICTURE CREDITS

Mark Ashbee 7 (btm), 138 (left), 153 (right top and btm), 169 (top) and 176

Trevor Edwards 37 (btm); Garrett Findlay 18; Priit Hõbemägi 224 (right); Sean Jackson 62 (top), 224 (left); Bo Oster Mortensen 24 (left)

Richard Murphy 2, 6-7 (top), 8, 20 (top), 21 (middle), 36 (top), 37 (top), 52, 53, 58-59, 59, 64 (top left), 66 (top), 67, 68 (top), 70 (top), 82 (btm), 82-3 (centre), 83 (btm & inset), 87, 92, 95 (btm), 100, 101 (btm), 102, 105, 108-9, 111 (top left), 113 (btm), 118, 121, 127 (top), 128-9, 131 (btm), 141, 142 (top right), 143, 144 (top), 145, 149, 150 (btm), 155, 156 (left), 161 (btm), 163 (btm), 170-1, 184 (top left), 184 (right), 185 (centre btm), 185 (right top), 188 (left top), 188 (centre top), 191 (left), 191 (centre btm), 191 (right), 192 (centre top), 193 (left btm), 194 (right), 195 (left), 197 (left), 197 (centre btm), 197 (right), 198 (right), 199 (centre), 200 (right), 201 (left top & btm), 202 (left), 202 (right btm), 203 (left top & btm), 205 (left top & btm), 206 (centre btm), 207 (centre top & btm)

Bernard van Giessen 6 (btm), 12, 15 (top), 16-17, 21 (top & btm), 22, 37 (middle), 38, 40 (top & middle), 50, 54-5, 57, 58 (top left), 61, 62 (btm), 64 (top right), 65, 66 (btm), 68 (btm), 68-9 (top), 72-3 (btm), 73 (right), 74 (btm), 75 (top right), 75 (btm), 77, 78, 79, 80 (top left and inset), 80-1 (top), 82 (top left), 85, 86, 88, 89, 90-1, 93, 95 (top, centre), 96, 97, 99, 106, 107, 110, 111 (top right lower), 113 (top left & right), 114 (btm), 115 (btm right), 122-23, 125, 126, 130 (btm), 130-31 (top), 131 (top), 132, 134, 135, 136-7, 139 (btm right), 142 (top left), 144 (btm), 147, 150 (top), 151, 152-3 (top), 153 (left), 156 (top right), 157, 158, 160, 162 (btm right), 163 (top & middle), 166, 167, 168, 169 (btm), 172, 173, 176, 179 (top left & right), 180 (top & btm), 182, 184 (btm left), 184 (centre), 185 (centre top), 186 (centre top), 187 (left top & btm), 188 (centre btm), 188 (right btm), 189 (centre btm), 189 (right), 190 (right bottom), 191 (centre top), 196 (left), 197 (centre top), 198 (top & btm), 200 (left), 202 (centre top & btm), 204 (left), 206 (right top), 206 (centre top), 209 (left), 209 (right)